Battle Notes for Wargamers

Battles with Model Soldiers
Colonial Small Wars
Wargames
Naval Wargames
Air Wargames
Advanced Wargames
Tackle Model Soldiers This Way
Wargames Campaigns
Handbook for Model Soldier Collectors
Military Modelling
All for a Shilling a Day
Bowmen of England
At Them With the Bayonet!

Battle Notes
for Wargamers

Donald Featherstone

David & Charles : Newton Abbot

0 7153 6310 7

Set in 11 on 13 point Times Roman
and printed in Great Britain
by Clarke Doble & Brendon Ltd Plymouth
for David & Charles (Holdings) Limited
South Devon House Newton Abbot Devon

Contents

List of Maps

Introduction

THE IDEA of refighting the famous battles of history, reproducing Waterloo, Gettysburg, Alamein, etc, on the wargames table, is a most attractive proposition. Unfortunately it is almost impossible to put it into practice with any degree of realism. A significant part of the history of the world, these battles involved thousands of soldiers fighting over vast areas of ground—at the Battle of Waterloo there were 100,000 French, 68,000 British, Dutch, Belgians and Germans and more than 100,000 Prussians. Even drastically scaling down this number, so that 100 men on the battlefield approximate to 1 man on the wargames table, gives ludicrous infantry battalions of about 7 or 8 men, and the wargamer will still need about 3,000 model soldiers! Waterloo was a relatively small battle for the number of troops involved, yet it stretched across a frontage of about 4 miles and, even with a vastly reduced ground scale, an unreasonably large wargames table would be required to reconstruct it. The same applies to Gettysburg, and, considering that Alamein was fought on a 38 mile front over 11 days, the almost insurmountable difficulties of recreating major battles can be appreciated.

However, military history abounds with conflicts that are highly suitable for reproduction on the wargames table—battles that involve small numbers, so that realistic scaling down is possible. This is important because, if the battle is to bear anything more than a titular resemblance to the original, the table-top armies must represent an accurate proportion of the original forces, both in numbers and types. It is possible in the battles described later for one man on the wargames table to equal as few as 5 men in real life, and some battles, such as Pork Chop Hill, can be fought on a man-for-man basis. The size and topographical features of the historical battlefield must lend

9

themselves to reproduction on the wargames table, and all those considered in this book are suitable in area and frontage for this purpose. Each battle also possesses tactical and human interests that make it worthy of simulation.

Table-top reconstructions of an historical battle must conform as closely as possible, to the events that occurred on the battlefield itself. It is pointless to construct a terrain resembling the battlefield of Maida, for example, and then let loose upon it a host of infantry, cavalry and artillery whose general milling about bears no relation to the original conflict. Every aspect of the battle has to be considered in its correct context, and in chronological order, so that it can be simulated and affected by the fluctuations and fortunes of war, without radically departing from the Military Possibilities of the day. The wargamer should consider what might have occurred in conjunction with what did occur.

It is doubtful whether wargames will ever give one profound military insight, but the wargamer may gain an understanding of the problems of the commanders in the field and a glimpse of the military thinking of the time by refighting each battle in the correct tactical manner, using the formations and weapons of the day. The purpose of this book is to discuss these factors and to suggest practical methods of simulation that will produce an accurate, realistic and enjoyable facsimile of the original battle. From the point of view of wargaming purely and simply, the battles described can be transported with magical ease backwards or forwards in time, so that the terrain and situation at Pharsalus in 48 BC, for instance, can be used by Wellington's seasoned Peninsular troops, or the rugged country around the Little Big Horn labelled 'A Tract of the North-West Frontier of India' and Custer replaced by one of Queen Victoria's commanders.

To refight any historical battle realistically, the terrain must closely resemble both in scale and appearance the area over which the original conflict raged, and the troops accurately represent the original forces. The most obvious manner of conducting the battle is for the troops to perform precisely the same manoeuvres as they did in the past, take the same percentage of losses and achieve the same success or failure. This is an historical exercise, not a wargame, and will only serve as a demonstration of what occurred during the real-life battle. It is the manner in which public demonstrations of the Battle of Hastings were carried out by the author during that town's 900th anniversary celebrations.

Another method is to follow the original course of events reasonably well, but allow some leeway, without too much imaginative stretch, for a reversed result. Too many liberties may not be taken, however, or, as we have said, the battle will become a wargame played for its own sake, lacking any precision. We must not allow the opposing table-top generals the advantage of hindsight, so that they can perform tactical manoeuvres far in advance of those known to their counterparts on the historical field. In a battle between Ancient Britons and Romans, for example, the commander of the former would have been most unlikely to throw his entire force forward in the characteristic headlong charge favoured by the Britons because he would know that it would be repulsed by the disciplined formation and tactics of the Romans; nor would he have conceived a tactical plan that included feint attacks, outflanking movements, feigned withdrawal to destroy Roman cohesion or holding back a reserve; his woad-covered Britons would never have obeyed orders with the discipline and steadiness of the Grenadier Guards. The wargamer should remember the tactical limitations of his troops, therefore, or we shall find, if the commander of the Ancient Britons is more conversant with table-top tactics than his Roman counterpart, that the semi-savage Britons are being handled in a tactical fashion superior to the well trained and highly disciplined Romans.

It is essential, therefore, that the tactics and formations of the original battle are reproduced and that all troops are obliged to conform to their known standard methods of fighting. For example, at Pharsalus the Roman legion formed up its 10 cohorts three-deep on a frontage of 2,000 ft, so allowing each legionary 6 ft of space in which to fight. On the flanks and in front the auxiliary slingers and archers skirmished and flung their missiles. Following the cardinal rule of dealing with an enemy on the defensive, Caesar's legions began their attack when 120 yd away, with the front line of cohorts moving forward at the march and then at the double until the first two ranks of each cohort were about 60 yd from the enemy. Then they launched themselves forward and hurled their pila at a range of about 20 paces; their comrades followed up and repeated the performance until the back lines of the cohorts had thrown their pila over the heads of their front ranks, who were now among the enemy, thrusting murderously with their short swords. They fought for exactly 15 minutes before being withdrawn and a fresh century or cohort thrown in. At the same time, the second line of cohorts moved forward, their front men preparing for their 60 yd run. The third line of cohorts was held in reserve to mop up the last of the enemy resistance or to cover a retreat if necessary. In reconstructing the battle of Pharsalus, therefore, the wargamer should employ these tactics.

It is important also to read varying accounts of a battle and the events leading up to it beforehand, because therein may lie any number of factors influencing its trend and pattern. Perhaps bad weather or muddy roads caused a significant proportion of one army to be late or to fail to arrive on the battlefield at all. Consideration of the campaign reveals the objective of the battle, such as the Black Prince attempting to join with an ally or to return home with the booty of the expedition, but being forced to fight at Poitiers. Armed with knowledge of a commander's historical intentions, one may better understand his tactical plans. One may vary the objective on the wargames table: for example, at the Little Big Horn Custer may try to fight his way back to join Reno instead of pressing forward to destroy an Indian force which, unknown to him, is overwhelmingly superior in numbers to his own.

Primed with background reading, the wargamer should now analyse the battle, seeking those moves that led to victory or defeat. Each phase should be considered carefully, while seeking possible alternatives to the historical trend of events. These alternatives will be called Military Possibilities, which may be defined by reference to Custer above. If he had joined Reno, what would have happened? Would their combined force have been able to hold out! You may find out on the wargames table.

The wargamer can enter the brains of Julius Caesar, the Black Prince, Marlborough, Sir Harry Smith, Lyon at Wilson's Creek, the ill-fated General George Custer, and Sir Paul Methuen at Modder River. He can win a VC with Newman at St Nazaire, and grapple with the problems of a brave lieutenant like Clemons on Pork Chop Hill.

Over and above the normal provisions of conventional wargames rules, certain factors must receive emphasis if an accurate and realistic simulation is to be achieved. Frequent references to these factors will be encountered in the various battle reconstructions that follow, and they are considered herewith in greater detail.

Military Possibilities

Military Possibilities are controlled and logical alternative courses of action that, had they been taken at the appropriate time during the battle, might well have caused a complete reversal of the result. Courses of action followed by opposing wargamers with the agreement of both, Military Possibilities are not excuses for indulging in whims and fancies nor for diverting events merely 'to see what happens'. In some cases the course of action indicated by a Military Possibility results in a more reasonable and credible result than occurred on the field of conflict.

Some Military Possibilities might depend on Luck, represented by the throw of a dice or the turn of a Chance Card, the simplest means of simulating the ebb and flow of war.

It must be decided whether Military Possibilities are radically to alter the historical course of the battle or be restricted to relatively minor aspects of it, when they may result in a more interesting 'twist' in tactics, or even influence the eventual outcome.

Numerous Military Possibilities are included in the simulation advice and suggestions that accompany each battle; such as at Cheriton where the young and impetuous Royalist cavalry commander, Sir Henry Bard, is made to resist his impulse to attack the Roundhead cavalry, so forcing Waller either to attack the Royalists in their strong defensive position or to withdraw from the field. Another allows the Coldstream Guards to ford the Riet River during the early stages of the battle of Modder River in 1899, when they could have taken the Boer position in reverse. Military Possibilities abound in all battles; the interest and colour they bring to the wargames table are proportionate to the ingenuity of the wargamer.

Ratings of Commanders

No wargamer likes to follow rigidly a course of action that history tells him will bring defeat, particularly when he has a marked numerical superiority, so, to save his face in such a situation, the original commander may be classified as 'average' or 'below average'.

If both forces are equal in strength, morale, equipment, position, manoeuvrability, etc, victory will almost certainly go to the better commander. In many of the battles described in this book a strong enemy force has been defeated by a numerically weaker army solely because its commander possessed outstanding tactical ability and was capable of inspiring his men to exceptional heights or, lacking either of those qualities, he just happened to be better than his opposing commander. Thus, just as history dictates that one commander was 'exceptional' while his opponent was only 'average' or 'below average', so this fact has to be reflected on the wargames table.

Taking the Battle of Poitiers as an example, the Black Prince must be classified as an 'above average' commander, as might be Warwick, Salisbury and even the Captal de Buch. The French King, the Dauphin and the Duke of Orleans must be classified as 'below average' commanders, while one of the two marshals, Clermont or Audrehen, can be an 'above average' commander.

The effect upon the battle of a commander's rating must be reflected by his troops, since those of an 'above average' commander will possess a higher standard of morale and better fighting qualities—represented by adding 1 (or any pre-decided score) to any dice affecting morale or fighting qualities. Conversely, a 'below average' commander would necessitate deducting 1 from dice scores, and the troops of an 'average' commander would remain unaffected.

An 'above average' commander could be given greater flexibility of movement and, within certain limits, the ability to undertake tactical movements outside his period (in other words, the wargamer is allowed to be himself). With certain reservations, the 'average' commander can be given the ability to alter characteristic fighting patterns, controlling and manoeuvring his force as suggested by their 'Style of Fighting' (explained in *Wargames through the Ages*). The 'below average' commander is given no such opportunity, and must control and manoeuvre his force without deviating from their 'Style of Fighting' whatsoever.

If a system is used where, at the start of the battle, orders have to be written for each army or group, the grading of commanders can be reflected by ruling that those armies or groups under 'below average' commanders must conform rigidly to their initial orders until they are disorganised by a forced reaction (such as a low morale rating). 'Average' commanders can write orders to carry through three moves of the game; at the conclusion of the third move fresh orders may be written if circumstances have not already altered the original instructions. 'Above average' commanders may write orders at the beginning of each move of the game.

An alternative is to allow all commanders, whatever their rating, to write their orders at the beginning of the game, but the lower the commander's grade, the greater number of moves he must take before he alters his orders, while the 'above average' commander may change his instructions much more rapidly.

With the exception of Julius Caesar at Pharsalus, there was no historically 'outstanding' leader present at any of the battles described in this book, so that the rating of commanders must be by 'local' comparison, and the reconstruction notes for each battle include a suggested rating for them. However, because an 'above average' rating is given to Lord George Murray at Prestonpans, it does not imply that he was a commander of the same calibre as Caesar himself. Nor should chauvinism cause American readers to feel aggrieved because George Custer is rated 'below average' at Little Big Horn—the rating is solely on his tactical ability and does not reflect upon his courage. In passing, it is interesting to note that most 'below average' commanders were men of great personal courage, such as Custer, Sir Hugh Gough or Lord Methuen – perhaps bravery was bestowed upon them to compensate for marked deficiencies in other qualities.

Chance Cards

Chance Cards introduce pleasant and unpleasant factors that materially affect aspects of a battle, perhaps even its result. They pose eventualities—tactical, physiological or psychological—and the commander drawing such a card has to take practical steps to carry out the instructions it bears.

Each historical battle could have its own set of cards designed to cover eventualities likely in such a situation. Closely allied to Military Possibilities, Chance Cards form the 'human' element that may affect the purely tactical aspects of a Military Possibility. Chance Cards can affect such aspects as de Buch's flank attack on the French at Poitiers by setting obstacles in his way (see p. 37). Machiavellian umpires often draw the countryside in such a manner as to present unexpected obstacles.

Every battle of military history abounds with situations which, in wargames, could give rise to the use of Chance Cards. The 15 engagements described in this book possess their share—the more obvious are listed but many more will spring to the mind of the imaginative wargamer.

Time Charts

Most of our battles include some 'surprise factor' that needs recording, so that a check can be kept, for example, on the anticipated time of arrival of a flanking force. A Time Chart programmes such vital factors beyond dispute. The Chart must include those manoeuvres whose timing is an important feature of the battle; each of them, or stages in their accomplishment, should be treated as a move in the game.

A Time Chart is vital in keeping check on off-table moves on a map, where different forces are moving along various routes or attempting outflanking movements that will bring troops on to the table-top battlefield at some intermediate stage in the conflict. It is almost impossible to retain control of such factors without a Time Chart.

Keeping in touch with detached portions of a force might require their commander, perhaps unaware of their exact location, sending messengers, whose progress must be recorded on the Time Chart. Thus, the messengers' exact time of arrival is known, and the unit to whom they are bringing orders cannot react until those orders are received. The non-arrival or delay of orders provides Military Possibilities that can realistically alter the course of a historical conflict.

As an example, consider the longest battle in the book—the CCF attack on Pork Chop Hill. This Time Chart will need to begin at about 2200 hours on 16 April 1953, and last through that night, and all through the following day and night, until after sunset on 18 April. On the other hand, at Prestonpans, scarcely more than 5 minutes elapsed from the first onslaught on Cope's army to the breaking of his entire front line. But the Time Chart must start when Cope took up position on the higher ground around Tranent, so that it gives full scope for charting the Highlanders' outflanking movement, and allows for their possible discovery and interception at the Riggonhead defile. Then follows Cope's realisation of the new threat and his redeployment to face it, the Highland onslaught and speedy rout of the royal army. In this case, the amount of time taken by each game-move will need to be short.

Simulation of Surprise

Every one of the battles described herein was affected by the element of surprise. While victory frequently goes to the big battalions, it can also be vastly influenced by shrewd tactical moves or the mere fluctuations of fortune. Surprise movements of bodies of troops, so that they suddenly arrive upon the rear or flank of the unsuspecting enemy (as at Prestonpans), are difficult to simulate on the wargames table, but such movements charted on a map enable forces to be manoeuvred so that the enemy is unaware of their intention, and even their existence, until contact is made.

The scale of 1 to 12 is used for all terrain maps in this book, so that a wargames table 8 ft × 5 ft appears 8 in × 5 in on the map. Draw a map 24 in long by 15 in wide, which will cover an area of 9 wargames tables, with the battleground taking up the middle oblong. Then, in the surrounding oblongs, continue topographical features such as hills, rivers, roads, etc. Cover the map with a pattern of inch squares, by drawing fine lines with a mapping pen on the surface of the map, or else by laying upon it a transparent plastic sheet upon which such squares have been marked.

Movement on this map is to the same scale as movement on the table, so that infantry moving 12 in on the wargames table will move 1 in on the map; this is most important, because it is the means by which 'off-the-table' movements are made. To illustrate the practical use of this map, combined with the Time Chart, let us take the Battle of Barnet. Oxford's force chased Hastings' men off the wargames table and on to the larger area of the map as far as the village of Barnet itself, moving at an agreed speed (according to the rules) over the specified distance. Time (in the form of game-moves) will be taken up by (a) the pursuit, (b) Oxford rallying his force and halting the pursuit, and (c) his force's return to the battlefield (the wargames table), where it strikes Montagu in the flank by mistake. During this period the battle in the centre oblong will have been proceeding. Oxford's movements are known to both wargames commanders, but a degree of uncertainty can be introduced by considering certain Military Possibilities. The distance of the pursuit can be varied, for instance, so that it will take Oxford a greater or lesser time to pursue Hastings, and bring him back to the main battle earlier or later than anticipated.

Surprise is difficult to simulate in the reconstruction of a historical battle, but if the conflict is to be accurately reproduced the surprise *must* appear. However, it can be tempered by Military Possibilities arising from local rules, morale factors or other means suggested in this section. Although all the troops are in full view on the table-top, a certain degree of tactical surprise and apparent concealment is made possible by each commander initially drawing up a plan of the tactics he intends to use, giving a broad outline of the role to be played by each unit of his army. Both commanders have 8 playing cards, 2 being aces. Before each move a card is drawn, and the commander may alter a unit's allotted role if he draws an ace, but, if not, he is committed to his original plan. This method can be used in conjunction with the ratings of commanders, simulating the decisiveness of De la Rey and the 'woolliness' of Methuen at Modder River by giving the British 1 ace and the Boers 3 aces per 8 cards.

The flank attack by the Gascon Captal de Buch at Poitiers causes certain practical problems on the wargames table. In 1356 the French had no idea that this attack was coming in on their left flank, whereas in our reconstruction they are fully aware of the daunting fact. The English commander may be given the choice of sending de Buch out to his right or to his left, so that the French, while being aware of the flank attack, will not know from which side they are threatened. But they must take steps to counter it by allocating men to watch the flanks of their column, and these men will not count in the main mêlée. Chance Cards and Military Possibilities may well cause the flanking force to be so delayed that they will never arrive at all or be too late to affect the result of the mêlée.

The countryside surrounding the area of the battle (in the centre oblong) can be drawn inaccurately on the map of the commander who is due to be surprised, so that he is unable to estimate the possibilities of an outflanking movement or the time it will take to reach him. On the other hand, the commander of the outflanking side can be given an inaccurate map, a Military Possibility that will affect his surprise move and give the original loser a slight chance of reversing the decision. If such steps are taken, it is advisable to have an umpire with an accurate map, for that will save a lot of arguments.

A reasonably successful method of surprise and concealment, again using maps that can be accurate or inaccurate as desired, requires the commander making the surprise move to work out, on a scaled map, the number of game-moves that it will take. He writes down

    (a) the unit or units involved (or total strength of force),
    (b) route from A to B (starting point of move to point of contact),
    (c) number of moves required.

If the force making the surprise move was visible in the battle, it will remain on the wargames table and will not move until the game-move in which it strikes home. If it was *not* visible to the enemy, it will move exclusively on the map, to be revealed and placed on the table when its presence was discovered. Using written movement orders (described in Chapter 2 of *Advanced Wargames*) and with a scaled map of the terrain, the commander perpetrating the surprise plots it for the required number of game-moves, writing instructions in progressive columns. On the completion of the requisite number of moves, the surprise force is disclosed by the wargamer controlling it. He must declare if he wishes to terminate his concealed move before its completion, when he must move his troops up to the point they have reached. Surprise is then considered to have been lost, even if his troops are now technically concealed. A suspicious enemy commander may challenge, but his suspicions will need to be reasonably precise, as to direction and intention —the advisability of using an umpire is stressed. In the event of something happening on the wargames table that interferes with the surprise move, such as an enemy force crossing or positioning itself on the line of march, the move will have to be revealed and a decision taken as to which side is more surprised. This eventuality could take the form of a Military Possibility.

Another method of simulating a concealed force requires each commander to have a set of 8 or 10 progressively numbered terrain maps, drawn on tracing paper. At the start of the battle each commander draws on his No 1 map blocks representing his troops at their agreed starting point and, from each block, draws an arrow indicating his first move in scaled move-distances. The umpire holds one map over the other to see whether or not any troops are within visual distance of each other or have come into contact. Each commander then places his No 2 over his No 1 map and draws blocks over the ends of the arrows indicating the first move, before drawing in the arrow representing his second move. The umpire checks both maps and marks on each the position of any enemy forces that can be seen. The procedure carries on with progressively numbered maps until a contact is made; then the troops are placed upon the wargames table, having approached each other as through the smoke of a battlefield.

Perhaps the simplest way of achieving surprise is for the host not to tell the visiting player the name of the battle he is fighting. For example, without mentioning the name of the battle the host could give the visitor the story of the events leading up to Cope's army forming up and facing south at Prestonpans, a course the visitor follows on the wargames table only to find, as did Cope, that the Highlanders have suddenly arrived on his left flank. An obvious snag to this procedure is that the host will not be able to take a leading part in the battle but may have to be umpire or hold a subordinate role under one or other of the commanders. If a 'host' is arranging a battle to be fought between two other wargamers, the relative insignificance of the battles described in this book may mean that one or both contestants have little idea of what occurred.

Morale
Morale concerns the discipline and confidence of troops, both collectively and individually, and was believed by Napoleon to be three times as important as physical factors. In at least 12 of the 15 battles described later the morale of the troops played a major part in deciding the outcome of the battle. Thus, there must be adequate simulations of this intangible factor, which causes men suddenly to break, or to rally and beat a force much larger than themselves, in spite of being attacked in flank or rear and having lost their officers. The Battle of Guilford Courthouse is an excellent example of such conduct.

The reasons behind a soldier's fear or exaltation have not changed since the beginnings of time. The hail of arrows from the English archers at Poitiers in 1356 aroused the same consternation and wavering among the French as did the grenades thrown by the CCF on Pork Chop Hill in 1953. Thus, there will be a basic resemblance in morale rules for all periods, though aspects peculiar to various ages must be considered.

A familiar wargames method of allowing a smaller force some chance of success is to use morale rules that cause their numerically superior but otherwise inferior opponents to break and run, often before or at minimum contact. This factor is suggested as the best method of simulating Sir Harry Smith's success at Aliwal.

Artillery
Modern guns are too long-ranged to allow them to be placed on the wargames table with the enemy upon whom they are firing. In the battles of Modder River, Gallipoli and Pork Chop Hill, for instance, artillery fire is simulated by 'off-table map-shoots', which are carried out by marking the guns on a map such as those already mentioned and then concealing a figure (an observer) on the wargames table to lay the guns on to the targets he can see. To register a nominated aiming point, the

observer must make a dice score of 5 or 6, which means that a hit is registered on that point (marked by a counter). Three such aiming points may be held at any one time. The observer may bring fire on to them unless the gun he is spotting for has moved (when they must be completely re-registered). The aiming points are also lost if an observer is killed. Guns may extend their target area by the observer nominating an aiming point and then altering the position of the 'Burst Pattern', providing a part of the pattern is still touching that point. A 'Burst Pattern' is a 6 in square of transparent plastic with four 2 in diameter circles numbered 1, 2, 3 and 4; the pattern bears a painted arrow which must point directly towards the firing gun. To simulate a shell burst, a pin is pushed through a centre hole in the 'Burst Pattern' into the 'hit' counter; the pattern laid, arrow directed towards gun, and a dice score of 1 to 4 indicates the destruction of anything in that specifically numbered circle. A dice score of 5 or 6 indicates the shell has buried itself in the ground without exploding.

Artillery firing without an observer requires the map of the table to be divided into a grid of 12 in × 12 in squares, each of them being again gridded into six 2 in squares, numbered 1–6. The firer nominates the large square, and dice thrown indicates the small square into the centre of which the 'Burst Pattern' pin is pushed. A further dice throw reveals the point of the shell burst. Details of 'off-table map-shoots' are given in *Advanced Wargames*.

1

The Battle of Pharsalus

9 August 48 BC

AT THIS stage of the Roman Civil War (50 to 44 BC) both armies were manoeuvring in Macedonia; after a reverse at Dyrrachium, Caesar had retreated 200 miles south-west into Thessaly, with Pompey's larger force cautiously following. Both armies camped on opposite sides of the Plain of Pharsalus, as Caesar regrouped and was reinforced by the troops of Domitius Calvinus while Pompey's army was strengthened by the legion of Metellus Pius Scipio; the men of both armies were irritable and pressing for a decisive action. On 9 August 48 BC (in modern dating), deciding to risk an attempt to overwhelm Caesar's numerically inferior force, Pompey formed line-of-battle on the plain between the opposing camps; his force totalled 60,000 infantry and 7,000 cavalry. With only 30,000 infantry and 1,000 cavalry, Caesar formed his tough and dedicated legions into the customary three lines, extending the intervals between cohorts to match the frontage of Pompey's army; his third line acted as a reserve. Caesar held back and took personal command of 6 cohorts (about 2,000 men) to cover his right rear and support his cavalry. His left flank rested securely on the steep bank of the Enipeus River.

Pompey's legionaries were commanded by Metellus Pius Scipio while his cavalry were led by Titus Labienus. Caesar's left wing was commanded by Mark Antony; his centre by Domitius Calvinus and his right wing by Publius Sulla.

At this time the normal Roman formation had the infantry in the centre and the cavalry on the wings to prevent the centre from being out flanked. Once the battle got under way and the enemy started to retreat, the cavalry moved forward and cut them down, so the main fighting was done by the infantry with the horsemen as a secondary force. Pharsalus was a notable exception because Pompey, having a 7 to 1 superiority in this arm, used his horsemen as shock troops against Caesar's right wing; otherwise Pompey was uncharacteristically lethargic and relied on a tactical plan based on his superior numbers and his strong cavalry force, placed on the left to make an outflanking movement.

As soon as his dispositions were completed, Caesar ordered his first two lines of infantry to advance and attack Pompey's motionless army; on his immediate right flank he posted his small force of cavalry. To steady his troops, who lacked battle experience, and calculating that it would make his flank attack more effective, Pompey ordered his troops to stand and await Caesar's attack. As the lines of infantry clashed together, Pompey launched his large force of horsemen against Caesar's cavalry. Although greatly outnumbered, the latter fought stubbornly and were only forced back by sheer weight of numbers.

It has been recorded that Caesar had trained a force of light infantrymen, one to each cavalryman, to work together in mutual protection so that the foot soldiers protected the horsemen who had been unseated and the horsemen protected the infantry when they were under pressure. Knowing that Pompey's horsemen were young and inexperienced, Caesar ordered his light infantry to thrust their javelins upwards into the faces of the horsemen, causing them to draw away in disorder. Although they were forced back, Caesar's cavalry played their part in pinning down the enemy's numerically superior force until, at the decisive moment, Caesar personally led his selected reserve of 6 cohorts against the flank of Pompey's horsemen.

The experienced legionaries scattered the surprised cavalry before them, then pushed on to slaughter some light infantry (archers and slingers) who were following them up, before

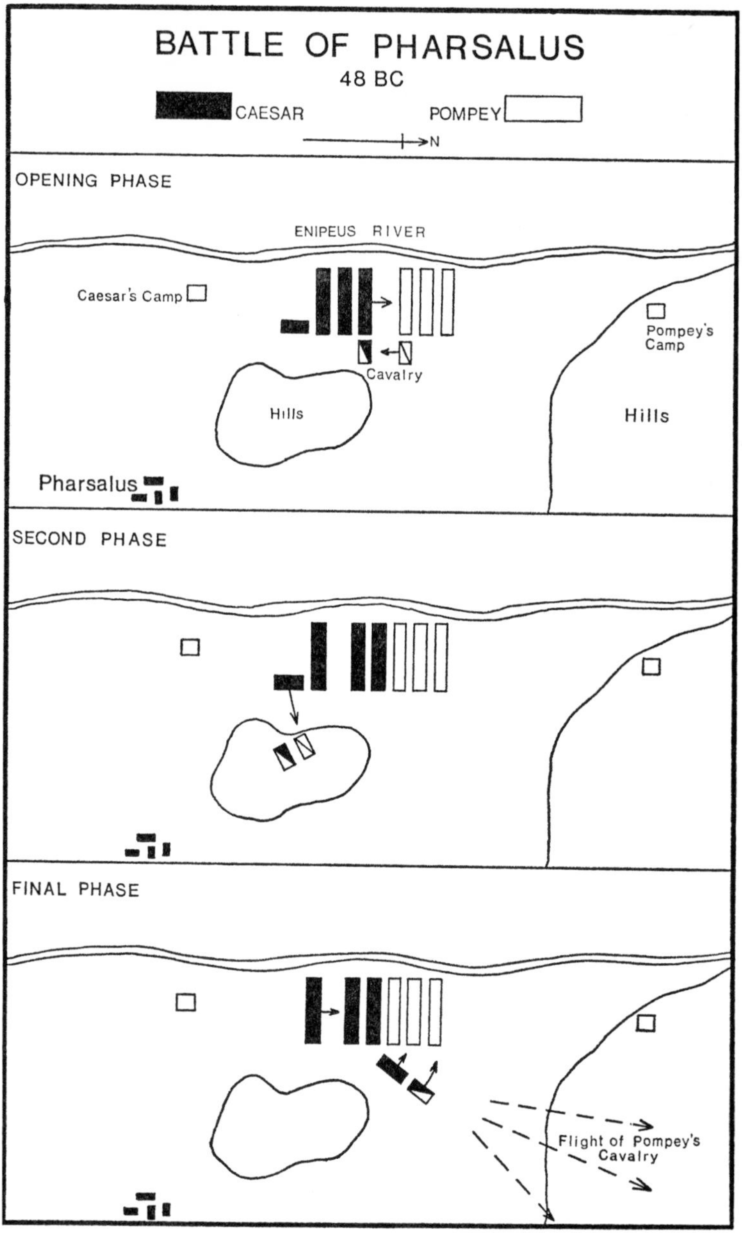

BATTLE OF PHARSALUS
48 BC

CAESAR   POMPEY

N

OPENING PHASE

ENIPEUS RIVER

Caesar's Camp

Pompey's Camp

Cavalry

Hills

Hills

Pharsalus

SECOND PHASE

FINAL PHASE

Flight of Pompey's Cavalry

turning against the left flank of Pompey's main army, which was heavily engaged with Caesar's two lines of legionaries. Leaving the 6 cohorts he had just led to victory, Caesar galloped to join his third line of infantry, to lead them through the intervals of the first two lines and smash into the front of the now exhausted Pompeian legions. Deserted by their cavalry, over-whelmingly assailed in front and flank by Caesar's legions, the Pompeian troops broke and fled.

Without halting his victorious troops, Caesar pursued the fleeing army to storm the enemy camp and, without letting his men stop to plunder, pursued the fugitives until they dispersed in all directions. Pompey fled and embarked on a vessel for Egypt, where he was murdered shortly afterwards.

In this decisive battle Caesar lost 200 legionaries and 30 centurions, together with about 2,000 wounded. Pompey lost 15,000 killed and wounded, and 24,000 of his men were taken prisoner.

The numbers involved here are larger than for any other battle in the book, but this is unimportant as long as the wargamer uses the number of cohorts employed in the historical battle (their table-top size and strength are immaterial). Because of the ready availability of Airfix Roman soldiers, this battle is most suitable for reproduction by groups of wargamers, each supplying his own Roman forces.

Rating of Commanders and Observations
Julius Caesar must of course be 'above average' while, on the day, it would seem that his experienced opponent Pompey should be classified as 'below average', as should his cavalry commander Titus Labienus. Of Caesar's subordinates—Mark Antony, Domitius Calvinus and Publius Sulla—the first named is said to have performed very creditably, and together they can be classified as 'average', as can Pompey's other subordinate, Metellus Pius Scipio.

The prime consideration in the reconstruction is to balance up the considerable numerical differences between the troops, by means of Caesar's superior rating and his actions, first with his 6 cohorts and then with his third line, and this balancing may be done by giving all his troops superior morale and fighting ability ratings. Their fighting ability is illustrated by the exploits of Caesar's infantry, who, though fewer in numbers, showed themselves to be tougher and more experienced than Pompey's. The Roman infantry method of fighting is described in the Introduction (p 12).

The battle will start without any manoeuvring, with Pompey's motionless formations being attacked by Caesar's first and second lines. By not moving forward, Pompey's formations of infantry failed to blunt the impact of Caesar's legions, giving the latter a bonus in the resulting mêlée. A Military Possibility can be evoked here by allowing Pompey's men, in fact, to move forward, though that may tip the odds too far against Caesar's numerically weaker forces.

Outnumbered 7 to 1 by Pompey's cavalry, Caesar's horsemen must have fought bravely. However, since few wargames rules will allow any force to stand against such disproportionate odds, several Military Possibilities can be used: Caesar's cavalry can charge forward, so blunting the force of the attack, and their light infantry can be allowed to send a shower of missiles at Pompey's onrushing cavalry, causing them casualties that will necessitate them taking a morale test before making contact. It may well be that one or more of their squadrons are turned back, so reducing the odds against Caesar's smaller force. Obviously such a large force would not all come into action at once, and their fighting frontage would match that of their lesser enemy. This would mean that a large proportion of their force stretched back beyond the mêlée, with a large extent of exposed flank, into which Caesar personally led his 6 cohorts, with devastating effect.

A Military Possibility here could allow the outermost ranks of Pompey's cavalry to see the oncoming legionaries and to turn and meet them. But this would so distort a vital aspect of the battle as to place it beyond realism. A practical means of simulating Caesar's surprise attack can be chosen from suggestions in the 'Surprise' section (p 18).

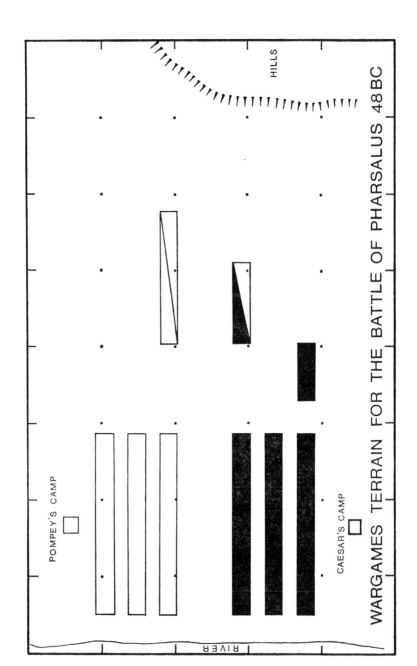

POMPEY'S CAMP

CAESAR'S CAMP

HILLS

RIVER

WARGAMES TERRAIN FOR THE BATTLE OF PHARSALUS 48BC

So important was Caesar's leading his third line of legionaries through the intervals of his first two lines to deliver a final blow to Pompey's infantry that it must be given special consideration, and can involve a number of Military Possibilities. For example, it could be decided that Caesar was too much involved with his infantry attack on the Pompeian cavalry to leave them, though as he did leave them, one might allow a drop in their morale after his departure. Another Military Possibility is to let the third line of infantry make their attack before Caesar joins them, and to see what happens to them without his personal leadership.

Construction of Terrain
This battle was fought on a perfectly flat plain, so that any wargames table or flat surface will suffice, though, for the sake of colour and interest, patches of scrub could be dotted over the terrain, with small groups of rocks, etc. To give greater manoeuvring space, the wargames map has been turned so that Caesar faces north and Pompey faces south, with the arena bordered on the west by the river and on the east by the hills.

2

The Battle of Poitiers

19 September 1356

THIS EARLY battle of the Hundred Years War was fought between an English army under the Black Prince and the army of King John of France.

The ridge occupied by the English army for about 1,000 yd was thick with scrub and undergrowth, and bounded by a hedge, with its left end falling away to a marsh and its right resting on open ground, strengthened by wagons, earthworks and trenches. Of the two gaps in the hedge, the upper gap was left open, but the lower was barricaded with stakes interlaced with vine branches. Between this ridge and the North Ridge, where the French Army massed, lay cultivated land, partly vines and partly fallow.

The English Army, about 6,000 strong, comprised 3,000 men-at-arms, with 2,000 archers and 1,000 sergeants, and was formed into Salisbury's division on the right, Warwick's on the left and the Prince's division, with a small body of mounted men, in reserve in rear. The men-at-arms were deployed into line, and solid wedges of archers were formed up on the flanks of each of the divisions, slightly in advance of them. The French army was about 20,000 strong, including 3,000 crossbowmen and two small contingents of 250 mounted men commanded by Marshals Clermont and Audrehen, and was formed into the Dauphin's division, the Duke of Orleans' division and that of the King. Remembering Crécy, the King of France dismounted his men-at-arms and shortened their lances to about 5 ft.

The two Marshals led their small mounted force forward

through the vineyard in a series of small columns, with Clermont's column bunching leftwards on the Nouaille Road while Audrehen followed the Gue de l'Homme track, each path bringing the two columns up against the twin gaps. The English archers on the left of their position kept up a galling fire and caused many casualties. Audrehen's men halted at the manned barricade, which their impatient leader jumped and was captured. Clermont's column passed through the unguarded gap, and swung right to be halted by Salisbury, who moved his line quickly forward, right up to the hedge, so closing the gap and preventing a flank attack on Warwick's division. After severe fighting the cavalry broke and fled; the English soldiers were rigidly restrained from pursuing.

The closely packed ranks of the advancing Dauphin's division were thrown into confusion and disorder as panic-stricken horses crashed through them, but got to grips when the English archers ran out of arrows. The hand-to-hand struggle surged backwards and forwards and, after Warwick's division was reinforced with the Prince's force, the Dauphin's men wavered and drew off in good order.

The division of the young Duke of Orleans was so shaken by the two repulsed attacks that, panic-stricken, it fled in scattered groups from the field. Seeing this, the King's Division of 10,000 men began slowly to advance, rolling forward in a glittering horde that alarmed the weary English. Scornfully dispelling their fears, the Black Prince mounted his force and sent the Gascon the Captal de Buch with 200 cavalry wide out to the right to hit the left flank of the trudging French column.

Led by the Black Prince, the mounted English men-at-arms rolled down the slopes towards the dip that lay between them and the North Ridge, with the mounted archers tacked on to the flanks and rear of their armoured comrades. Seeing the sudden avalanche of men and horses cascading down upon them, the advancing French division stopped, so that the rear ranks piled up on those in front, while others shambled from the field in panic, and before they could assume a defensive forma-

c

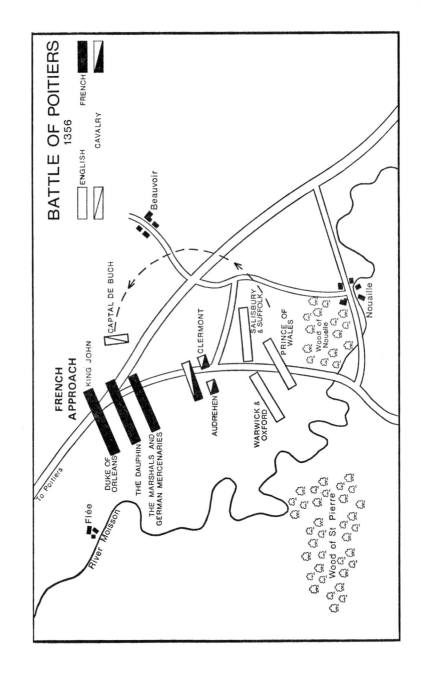

BATTLE OF POITIERS
1356

ENGLISH
FRENCH
CAVALRY

FRENCH APPROACH

KING JOHN

CAPTAL DE BUCH

Beauvoir

To Poitiers

Flée

River Moisson

DUKE OF ORLEANS

THE DAUPHIN

THE MARSHALS AND GERMAN MERCENARIES

CLERMONT

AUDREHEN

SALISBURY & SUFFOLK

PRINCE OF WALES

WARWICK & OXFORD

Wood of Nouaille

Nouaille

Wood of St Pierre

tion, the English horsemen crashed into them with a fierce shock that tumbled men and horses to the ground.

In the hard and bloody conflict the outnumbered English forced their way forward yard by yard in a mêlée that surged back and forth. No one saw the Captal de Buch's small body of cavalry coming in from the flank to drive deep into the King's division. The great French column, attacked on two sides, slowly disintegrated as men lurched from the field, until the King of France, realising that all was lost, surrendered. Slowly the battle burned itself out, with the triumphant English pursuing the fleeing French as far as the very walls of Poitiers.

The French casualties were approximately 2,500 killed, some 2,000 captured plus about 4,000 wounded. The English appear to have got off very lightly.

At a figure-scale of 20 to 1, the English force will consist of 200 dismounted men-at-arms and 100 archers, plus 10 horsemen, and the French 500 men on foot plus 25 cavalry, but the war-game will only require sufficient men-at-arms to form the Dauphin's initial attacking force of 250 men (representing 5,000) who, when dispersed, can become half the King's force of 500 men (representing 10,000). The Duke of Orleans' column of 5,000 never got into the battle.

In a wargame it is unlikely that the scaled-down force of 25 mounted French men-at-arms who attacked under the two Marshals will be very effective against 300 English in position, plus the massed fire from their archers. To simulate the battle, the cavalry must attack, but Military Possibilities can make their venture more than a suicidal mission. The vineyard can be sufficiently high to mask them from archery fire for most or all of their charge; the height of the vines could produce surprise by the horsemen bursting upon the English (particularly at the point where the gap in the hedge is not barricaded). At that stage the English must check their morale. If the English archers move into the marsh during the charge, their firing time will be taken up.

The Dauphin's column of 5,000 strong (scaled down to 250 dismounted men-at-arms) will attack next—the French *must* attack piecemeal as in 1356—and the trampled vines will cause it to split up into small columns to attack all along the English line. The French must check their state of morale, which will be affected by the retreating cavalry crashing through them plus losses from archery fire. If it is adequate, the English will test their morale to see if they stand—if they flee, the similarity between our reconstruction and the battle itself vanishes with them; but most wargames rules allow a force in position, aided by defensive fire, to repel an attacking force of approximately equal size.

Assuming that the Dauphin's column is repelled as in 1356, there may be a Military Possibility to decide if the English pursue one or both of the defeated French columns and so diminish their numbers. This can be decided by throwing two dice: a total of 4 or less means that both columns are pursued, a total of 5 that the Dauphin's column is pursued, a total of 6 that Orleans' column is pursued, and any total of 7 or over that no pursuit takes place.

If a pursuit is decided on, throw three dice to settle the numbers of pursuers—the combined total being the percentage of the remaining strength of the English force. For example, if the English have 200 men remaining on the table and the dice total is 15, then 15 per cent of their force (30 men) will pursue.

The North Ridge is crowned by the lance pennants of King John's host of 10,000 (scaled down to 500 dismounted men-at-arms). The morale of both French and English must now be considered. Reluctant to attack, the French morale might be low, but the English might suffer a temporary lapse in morale at the thought of attacking such a large force of the enemy. Nothing occurs to raise Franch morale, whereas English morale quickly rises because of the inspiring personality of the Black Prince. The English dismounted men-at-arms are all replaced by cavalry figures who charge down the hill towards the oncoming dismounted French. The Captal de Buch leads 200 horsemen (scaled down to 10) out in a wide arc to the right, coming in on the French flank, moving off the table (at scaled rates) on to a map of the countryside immediately to the right of the English position, and coming back on the table when they reach the French column. Timing is of primary importance here; de Buch should move *before* the Black Prince. Military Possibility might make them move off simultaneously, besides presenting delaying obstacles to de Buch's progress—a sunken road, a small ravine, thick hedges to hinder his progress or a stretch of marsh that looks like lush green grass. A well designed group of Chance Cards will be very useful at this stage in the battle.

In wargames it has to be decided whether the infantry's morale will allow them to stand against horsemen and, should they stand, whether the cavalry, shaken by their losses from missile fire, suffer in their morale. There were some 3,000 (scaled down to 150) mercenary crossbowmen with the French King's force, and few sets of wargames rules will allow the English to charge home in the face of crossbow fire. It is extremely unlikely in an army of this period, however, that the crossbowmen were deployed in front like skirmishers; they were far more likely to be bunched somewhere in the middle of the column or in the rear. Therefore, probably no more than half of them (possibly far less) will fire upon the oncoming English cavalry. A morale test will reveal whether they hold their ground and fire or let go at long range and flee.

A sprawling hand-to-hand combat will end the battle one way or the other. In 1356 the French were taken unawares by de Buch's flank attack, but that is unlikely in our wargame. A Military Possibility can allow the English commander to send de Buch out to his right or left; the French will then be forced to allocate a force (which will not count in the main mêlée) to watch their flanks. On the other hand, it is possible that Chance Cards, etc, may cause the flanking force to be delayed or never to arrive at all.

Rating of Commanders
The Black Prince must undoubtedly be classified 'above average' as must Warwick and Salisbury and even the Captal de Buch. The French King, the Dauphin and the Duke of Orleans must

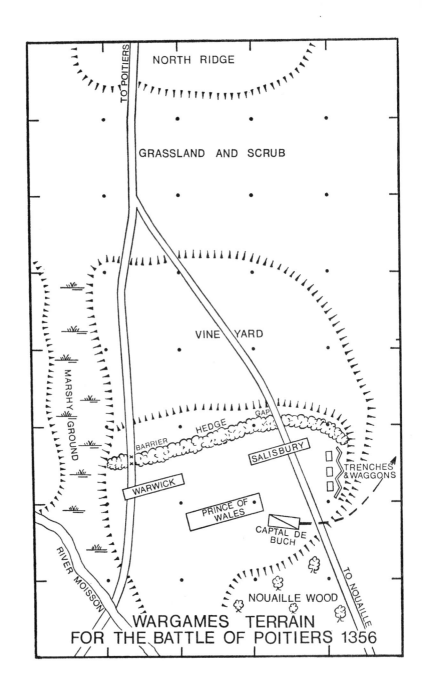

TO POITIERS

NORTH RIDGE

GRASSLAND AND SCRUB

VINE YARD

MARSHY GROUND

HEDGE GAP

BARRIER

SALISBURY

WARWICK

TRENCHES & WAGGONS

PRINCE OF WALES

CAPTAL DE BUCH

RIVER MOISSON

TO NOUAILLE

NOUAILLE WOOD

WARGAMES TERRAIN
FOR THE BATTLE OF POITIERS 1356

be classified as 'below average', while one of the two Marshals, Clermont or Audrehen, can be 'above average'.

Construction of Terrain

Fight the battle lengthways rather than across the table. The ridge and its approaches covers about 9 sq ft, for the first stage of the battle with the horsemen and the Dauphin's column. Extending $4\frac{1}{2}$ ft forward is an area of about 16 sq ft for the mêlée with the King's column. This allows sufficient space for the crossbowmen to fire on the mounted English attack on the French King's forces.

The ridge on which the English formed up will cover approximately the lower half of the table, allowing a gradual slope up through the vineyard for the initial cavalry attack and the advance of the Dauphin's column, besides allowing the English archers to get off a reasonable number of shots at the advancing French. This is most important, because it is the firepower of the English archer which really makes possible this battle between two forces of such vastly different strengths.

This terrain may easily be constructed of planks of wood, slabs of polystyrene or books on the table-top, covered first by sheets of newspaper or blankets and then by a green cloth to fall realistically into the right contours. The tracks can either be coloured in or made of strips of suitably coloured adhesive paper. The scrub and vines are simulated by lichen moss.

3
The Battle of Barnet
14 April 1471

THIS BATTLE was fought during the Wars of the Roses between the Yorkists under Edward IV and the Lancastrians under the Earl of Warwick.

Returning from exile in March 1471, Edward IV landed in Yorkshire and marched south. Warwick's opposing forces manoeuvred until the night of Saturday, 13 April, when they positioned themselves near Barnet, the men lying down in their ranks behind a long line of hedges. During the night the Yorkist army reached Barnet, Edward marching them up to Warwick's position in the dark (a rare accomplishment in those days) and drawing them up as best he could. Edward was a good soldier, and realised the necessity of forcing his enemies to fight before they could gather their full strength against him.

Edward drew his troops up in battle order facing Warwick's army on the lower ground beneath the road. In the darkness, Edward's right wing extended beyond Warwick's left, and the latter's right outflanked Edward's left. Realising the enemy were at hand, Warwick opened a fruitless cannonade from his guns posted on his right.

The morning of Easter Sunday, 14 April, brought with it heavy fog, so that the armies groped blindly forward. Edward formed his army up in three bodies, commanding the centre himself, and giving his younger brother, Richard, Duke of Gloucester, command of the right wing and Lord Hastings command of the left. The Lancastrians were formed up in four groups, with Warwick himself leading the left wing, Exeter

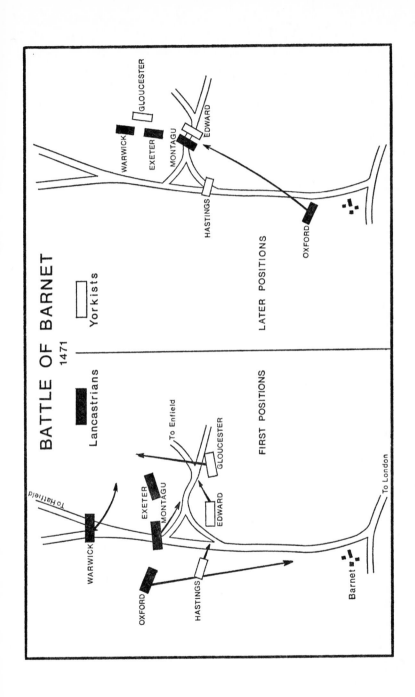

BATTLE OF BARNET
1471

Lancastrians ■

Yorkists □

FIRST POSITIONS

To Hatfield
WARWICK
EXETER
MONTAGU
OXFORD
HASTINGS
To Enfield
GLOUCESTER
EDWARD
Barnet
To London

LATER POSITIONS

WARWICK
GLOUCESTER
EXETER
MONTAGU
EDWARD
HASTINGS
OXFORD

on his right, then Montagu, and finally Oxford commanding the right wing. The Lancastrians had 15,000 men and the Yorkists 10,000.

Both armies are known to have had artillery, but Warwick's was the better served. The foot soldiers were equipped with new weapons, for both Edward and Warwick had hired bands of Burgundian hand-gun men. Apart from these professionals, the armies were formed of nobles and their retainers; during the Wars of the Roses the nobles frequently changed their allegiance, so that men were now fighting alongside each other who, in past days, had been bitterly opposed; there were doubt and suspicion of treachery in the air.

The battle began in thick fog between 4 and 5 am, with both sides advancing slowly and cautiously, Edward's army having to ascend rising ground. At no time during the day was visibility more than about 20 yd, and as men loomed out of the mist, archers and cannon opened fire. Edward and Montagu were the first to find each other, and rushed in to engage at close quarters. On the Yorkist right Gloucester suddenly came upon Exeter's left flank, which turned and faced him, besides bringing Warwick in on its left, so that Gloucester was engaging both groups. On the Lancastrian right Oxford, an efficient soldier, hit Hastings' left flank, causing his force to break in sudden rout; the broken remnants were chased by cavalry down the road as far as Barnet itself.

Unaware of these events, Warwick could not take advantage of the success of his right wing, so Oxford's pursuit had no effect upon the rest of the battle. Unaware that his left was in the air, Edward continued pressing Montagu hard and getting the better of his West Country bowmen and billmen. Exeter, with his men well in hand, appears to have given some assistance to Montagu's hard-pressed force. Montagu had told his men to keep in line with Exeter's, who, to meet Gloucester's attack had turned to face eastwards; both armies had gradually turned anticlockwise until they were at 90° to their starting positions.

Oxford halted and reformed his men, to return to where the Yorkists should have been, and came upon the exposed flank of a large body of men looming out of the mist. This force, which was actually Montagu's Lancastrians, mistook Oxford's De Vere badge of a star for Edward's sun device, and fired volleys of arrows as Oxford's men charged into them. Suddenly recognising each other, both sides raised the cry of treason, always common in the Wars of the Roses, and many fled towards the Yorkists, who simply cut them down. There was general confusion.

Oxford, convinced that he had been betrayed, left the field. The cry of treason passed from man to man down the already hard-pressed centre of the Lancastrian army; Montagu and his men found themselves under attack from two sides and Montagu himself was killed. Pushing steadily forward, Edward then fell on Exeter, who fled the field followed by his men. Only Warwick's wing was left to maintain the combat, but, seeing that all was lost, he left the field, to be speedily overtaken and slain. Edward ordered 'no quarter', and estimates of the Lancastrian dead range from 4,000 to 10,000.

This battle is best reconstructed as three separate actions, perhaps in different rooms: (1) the battle between Edward and Montagu, (2) Gloucester against Exeter and Warwick, and (3) Oxford's rout of Hastings. The division into three actions simulates the confusion caused by the fog. When Oxford's returning troops blunder into Montagu's flank, all 6 wargamers could join together on one large table to finish off the battle.

Rating of Commanders and Observations
Edward should be classified as 'above average'; Fortescue writes that, despite the manner in which the fog became his ally, victory was due to Edward's promptness and his rapidity of decision, which stamped him as a soldier far in advance of his time. Warwick, who was also a very fine soldier, did not show up at his best in this battle—he should be classed as 'average', together with Montagu, Exeter and Gloucester. Oxford could be classified as 'above average', since his rout of Hastings was a well conducted affair. He might be credited with extra morale-

power to rally his soldiers and bring them back to the battle, since his return is a vital aspect of the Battle of Barnet.

The element of surprise here will be far from easy to simulate, and it is suggested that all formations move on the map, using tracings. Only when two forces moving on the map appear within a charge-move distance of each other should they be placed on the wargames table.

Edward's night march before the battle, including his final deployment, could be marked on a map, so that uncertainty will creep into the positioning of his forces. This may preclude Oxford from outflanking Hastings, and Gloucester from pinning down both Warwick and Exeter. Military Possibilities have to be considered in these cases.

The shaky loyalties of the troops at Barnet should be exemplified in the morale rules, which should be formulated to allow formations to break and rally easily; they may return to the fray somewhat timidly.

Simulating the difficulties of fighting in a fog can be handled by an 'Obscurity Factor', using 'mild' Chance Cards bearing relatively marginal instructions such as 'Unit (formation) moves 3 in forward . . . or back . . . or half left or right . . . or back-wards', each alternative being marked on a separate card. There are innumerable minor manoeuvres that will simulate the uncertainty of men moving in the face of an unseen enemy.

Oxford's rout of Hastings abounds with Military Possibilities. For the sake of realism, Hastings must be initially chased off, otherwise Oxford will not be able to return and create the confusion that led to the Lancastrian defeat. Some local rule must ensure that Hastings is hit in the flank and goes—perhaps Oxford's force could be made stronger than that of Hastings. Another Military Possibility could be to allow Hastings to make a semi-fighting withdrawal, delaying Oxford's return to the field. Anyhow, it has to be decided how far Oxford's men chase and how long it takes him to get them under control—this is largely a question of their morale. Military Possibilities could delay that regrouping so that the three forces of Warwick, Exeter and

Montagu might have the time to defeat those of Edward and Gloucester.

Another vital aspect of the battle was the manner in which the forces wheeled, causing Oxford's men on their return to hit Montagu in flank. This can be worked by assuming that Edward and Montagu (who made the first contact) edged forward towards each other to meet on or south of the cross-roads. As Gloucester was able to hold the combined forces of Exeter and Warwick, it is reasonable to assume that initially he might force Exeter back and, as Warwick moves forward, both end up facing east, presenting a united force against Gloucester.

Montagu's archers fired upon Oxford's men as they loomed up out of the mist, but there is no mist over our table-top battlefield nor will any wargamer fire upon his own men. By giving Montagu's archers some uncontrolled reaction (as described in the Ancient and Mediaeval Rules of the Wargames Research Group), one could force them to fire, or the matter could be handled by judiciously worded Chance Cards.

If neither wargamer is informed that it is the Battle of Barnet he is fighting, the wargamer representing Montagu may be told that a force is coming up fast on his flank (on the map) and will be on the table during the next move, coming within bow range of his ranks. He can then be given the options of holding his fire to see if they are friends or firing as soon as they appear. Oxford's men, coming on to the table from the map, may also be informed that they will be arriving on the flank of a force at the same time as they are placed in position on the table, and that this force may fire upon them.

Rules must be formulated to cover the firing of artillery and hand-guns, and to represent their undoubted effect upon morale. To cover the deaths of Warwick and Montagu, it is suggested that 'single combat' be built into the game, detailed instructions for which are given in *Advanced Wargames*.

Construction of Terrain

The top three-quarters of the field should form a flat plateau,

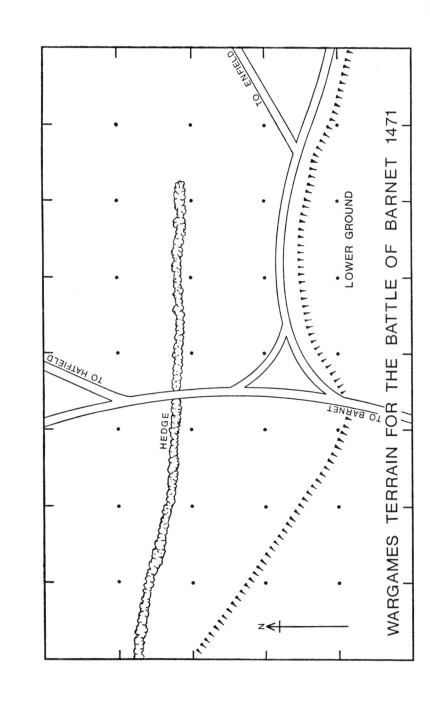

WARGAMES TERRAIN FOR THE BATTLE OF BARNET 1471

TO ENFIELD

TO HATFIELD

TO BARNET

LOWER GROUND

HEDGE

N

with the lower or southern part rising up to it (it was here that Edward formed up during the night). Although they play no part in the battle, the tracks shown on historical maps could be included on the table.

4

The Battle of Cheriton

29 March 1644

THIS BATTLE was fought during the English Civil War between a Royalist army under Lord Hopton and a Parliamentary force commanded by General Sir William Waller.

On 28 March 1644 the Royalist army camped on a horseshoe-shaped ridge, their guns placed to cover Cheriton Wood on their left. Forward in a little wood on a parallel ridge to the south was a detachment of 1,000 musketeers and 500 horse under Sir George Lisle. Next morning, 2 hours after sunrise, when the thick mist lifted, masses of Roundhead cavalry were seen deploying on the southern slopes of the ridge where Lisle was posted. Dangerously exposed, his force, covered by cavalry, had to retire. Parliamentary forces then moved forward and occupied Lisle's former position. Waller sent a force of cavalry down into the valley in front of his guns and 9 regiments of infantry, each drawn up 6 deep, with 5 regiments forward and 4 behind, lined the ridge behind hedgerows from Cheriton Wood towards the village of Cheriton.

Commanded by Lords Hopton on the left and Forth on the right, the Royalist army of 3,500 infantry, 2,500 cavalry and 10 guns, was formed up with infantry in the centre and cavalry on wings. Facing Cheriton Wood, Hopton's infantry were drawn up on reverse slopes. The Royalist army included the Queen's Regiment of cavalry, which had many Frenchmen in its ranks, and some redcoated Irish infantry regiments. The Parliamentary Army consisted of 6,000 infantry, 4,000 cavalry and 15 guns; the infantry included White and Yellow Regiments of the London Brigade and Horse and Foot from Kent, and the large cavalry force under Sir William Balfour included Hazelrig's cuirassiers, known as the 'Lobsters' because of their iron armour. A useful addition was Colonel Norton and his troop of Hambledon Boys, who knew the countryside around Cheriton.

About 1,000 men of the London Brigade went forward to capture Cheriton Wood, where hand-to-hand fighting took place with Sir George Lisle's men. Hopton had forseen this attack and planted some drakes (field-pieces) on the high ground north-east of the wood; when the London Brigade surged out from the trees, they ran into the point-blank fire of these guns, which forced them to retreat to the shelter of the wood. Royalist musketeers who followed them were repulsed, but Royalist infantry outflanked the Roundheads on the east of the wood and threw the London Brigade back in disorder. The Royalists then reoccupied the wood.

Waller had posted his Horse in such a position that if the Royalist cavalry approached they had to come down a lane and could only deploy one regiment at a time. But Hopton, in such a strong position that it was best to stand fast and make Waller decide whether to attack him or withdraw from the field, did not intend to take any cavalry action.

However, Sir Henry Bard, a young and impetuous cavalry commander, could not resist attacking the Roundhead cavalry whom he saw drawn up in front of him. His regiment charged down the lane towards them, taking fire from musketeers posted behind hedges and in coppices. Bard was killed and his horsemen overwhelmed, but troop after troop of Royalist cavalry poured down the lane after him, to be defeated in detail by the Parliamentary horse and musketeers as they emerged on to the open ground. Before long all the Royalist cavalry on the right had been defeated and Hopton's flank was open to attack; so, to retrieve the situation, the infantry were sent down and, meeting the Parliamentary foot soldiers, they came to 'push of pike' with both sides fighting stoutly. Eventually Hazelrig's Lobsters swept round behind the Royalist cavalry and

D

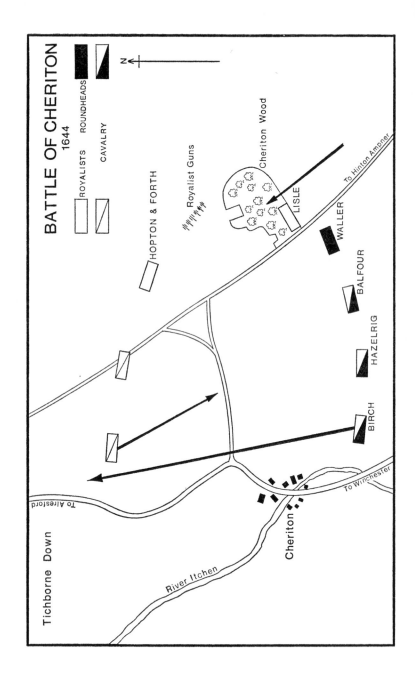

BATTLE OF CHERITON
1644

ROYALISTS   ROUNDHEADS

CAVALRY

N

HOPTON & FORTH

Royalist Guns

Cheriton Wood

LISLE

WALLER

BALFOUR

HAZELRIG

BIRCH

To Hinton Ampner

To Winchester

Cheriton

River Itchen

To Alresford

Tichborne Down

into the infantry, driving them back in disorder. Before that, a force of Parliamentary musketeers came out from their cover and fell upon the flank of the Royalist cavalry so putting the final touch to their discomfiture.

Another Parliamentary infantry attack on Cheriton Woods drove out the Royalist infantry, so that the whole Royalist line fell back on its original position while the last of Hopton's cavalry sealed off the end of the lane. The Royalist force withdrew during the night to Basing. Casualties were 900 killed and wounded in the Parliamentary army and 1,400 in the Royalist.

Rating of Commanders and Observations

Forth, Hopton, Waller and Balfour are 'average'; Bard and Lisle 'below average'.

The initial deployment of both forces in the mist can be simulated by moving them on maps, bringing them to the table when it is considered that the mist has lifted and they can see each other. Lisle's withdrawal during the opening stage of the battle could be subject to a Military Possibility, so that he fights instead of withdrawing, although the degree to which this is carried out could materially affect the later accuracy of the reconstruction. Hopton's infantry, drawn up on the reverse slopes facing Cheriton Wood, were not visible to the Parliamentarians—this could be simulated by marking them on the map but not placing them on the table.

About 1,000 musketeers of the London Brigade chased the same number of Royalist musketeers out of Cheriton Wood; this will be an even struggle on the wargames table and, without the aid of any Military Possibility, the Royalists might well throw back the London Brigade. To retain realism, therefore, the latter could be given a bonus in both morale and fighting ability, so as to ensure that they follow the historical precedent of capturing the wood. When the London Brigade emerged from the wood they ran into point-blank artillery fire, which caused their morale to falter; in 1644 they recoiled, and since it is desirable that they do so in the reconstruction, 'local' rules might be required. (On the other hand, Military Possibility might prevent the London Brigade from being beaten back, so that they attack the guns, perhaps capturing them.) The morale of the Parliamentary musketeers must rise when back in the shelter of the trees, because again they defeat their Royalist pursuers; but then more Royalists outflank them east of the wood and drive them out in disorder. A Military Possibility might enable the London Brigade to stand, so that throughout the course of the battle Cheriton Wood is held by Parliament.

At this stage Hopton, with his smaller army, wished to stand on the defensive and await Waller's attack; a Military Possibility might prevent Bard from charging down the lane at the head of his cavalry and precipitating the defeat of the Royalist horsemen at the hands of Parliamentary cavalry. But this is inaccurate and Military Possibility should be heavily weighed so that Bard *does* go. However, if he does not charge, Waller will either have to come forward and attack Hopton or withdraw from the field, leaving it in the possession of the Royalists. A Military Possibility could determine whether the rest of the cavalry follow Bard, how many of them do so, whether they go down at such intervals as to support each other or whether the massed Parliamentary cavalry at the foot of the lane is given sufficient time to destroy each squadron as it emerges. This might be the place for some sort of uncontrolled attack (as provided for in the Wargames Research Group's Ancient Rules). Another Military Possibility could allow the Royalist cavalry to hold their own or even be victorious—but this is so historically unrealistic as to destroy any semblance of a true reconstruction of the battle.

In the 'push of pike' that followed the Royalists held their own until Parliamentary cavalry burst through the Royalist horsemen and got round in their rear. Military Possibilities here could include a Royalist victory in the infantry battle or some factor preventing the Parliamentary cavalry from getting round behind the Royalist foot. Both these Military Possibilities are feasible in a battle that was never a runaway victory for Parliament. However, if the reconstruction follows history, the

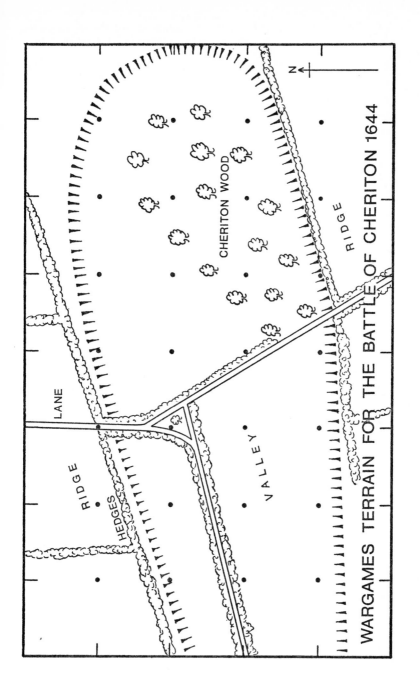

WARGAMES TERRAIN FOR THE BATTLE OF CHERITON 1644

CHERITON WOOD

RIDGE

LANE

RIDGE

HEDGES

VALLEY

N

battle finishes with a Parliamentary infantry attack on Cheriton Wood, forcing the Royalists to fall back to their original position, with the remnants of Hopton's cavalry sealing off the end of the lane. This allows an opportunity for an interesting rearguard action, without any Military Possibilities.

Construction of Terrain
The bulk of the action took place in and around Cheriton Wood and in the valley to the left of the lane that bisects the terrain. Both armies were positioned behind hedges on rising ground at the north and south ends of the table. The entire north and south quarters of the table need to be raised by slabs of polystyrene, planks, or any other means, to a height of about 4 in and then covered with a cloth depressed to form the valley that runs across the middle of the table. Cheriton Wood need not be as dense as shown on the map, but can be a sheet of green-painted hardboard with a few symbolic trees dotted around its edges. The hedges play an essential part not only because they deterred attack when lined with musketeers and pikemen but because hedges bordered the narrow lane and channelled the Royalist cavalry, squadron by squadron, down into the valley below.

5

The Battle of Wynendael

28 September 1708

THIS LITTLE known but typical small conflict, which took place during Marlborough's campaign in the Low Countries, is ideal for reconstruction. On 27 September 1708 a convoy of 700 wagons, escorted by 12 battalions of infantry, set out from Ostend for Lille, which Marlborough was besieging. To supply further protection to the convoy, General Webb marched with another 12 battalions to Thourout, while Cadogan took 26 squadrons of cavalry to Roulers, two towns on the convoy's route. Vendôme, the French Commander, had sent Count de la Mothe at the head of a substantial force to intercept the convoy.

On the following morning Count Lottum, sent with 150 horse by Cadogan to meet the convoy, reported that he had struck a strong French force at Ichtegem, 2 miles beyond Wynendael and 4 miles from Thourout on the Ostend road. Webb collected all his available infantry battalions and set off, with Lottum's squadron patrolling ahead, and, emerging from a wood-bordered defile on to a heathy plain, his cavalry came upon de la Mothe's advancing columns. The cavalry skirmished as they slowly retired, giving Webb time to form his men into two lines across the entrance to the defile, battalion by battalion as they marched up, his right resting on the Castle of Wynendael. Prussian and Dutch infantry together with Grenadiers were posted in the woods on either side.

Webb's force, totalling 6,000 men, was formed of Dutch, Prussian and Hanoverian troops, including Orkeney's Royal

Scots, Preston's Cameronians, and Collyer's and Murray's Scots-Dutch. He had no artillery. The French force consisted of about 24,000 men, made up of 60 squadrons of cavalry, 34 battalions of infantry and at least 20 guns.

Lottum's cavalry concluded their skirmishing and retired through their own infantry. Now a battery of 19 French guns opened a cannonade upon the closely formed Allied ranks that lasted for nearly 2 hours, but it had surprisingly little effect, for Webb had ordered his men to lie down. Then 4 lines of French infantry, supported by cavalry and dragoons, came forward and entered the defile between the woods, their flanks brushing the shrubbery. Suddenly, at a few yards' range, volleys of musketry poured out from both sides of the defile, causing the French flanks to recoil and reel back on their centre. Dragoons came forward in support and the infantry rallied, pressing forward so vigorously that they broke through two battalions of the first line. The gap was quickly filled from the second rank, however, and the French infantry were forced back. Eight lines of French foot then marched up and clouds of smoke filled the air as repeated musketry volleys assailed them from front, flank and rear until, in spite of the entreaties and blows of their officers, they broke and fled. The Allied infantry, advancing by platoons 'as if they were on exercise', fired volley after volley into them as they retired. Cadogan hastened up with a few squadrons of cavalry and considered charging the retreating enemy, but, realising that he was not strong enough to do so with any certain chance of success, stood off.

In a complete state of confusion the retreating French left more than 3,000 casualties on the ground and all their guns, which were recovered on the following day. Webb lost about 920 officers and men.

While the action was taking place, the convoy of wagons had been rerouted along a road to the rear of the wood, eventually arriving without loss 2 days later at Marlborough's camp.

The effects of this remarkable little action greatly relieved

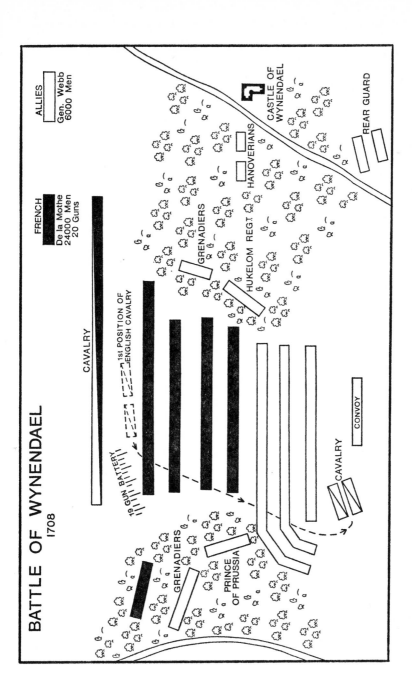

BATTLE OF WYNENDAEL
1708

ALLIES
Gen. Webb
6000 Men

FRENCH
De la Mothe
24000 Men
20 Guns

CAVALRY

1st POSITION OF
ENGLISH CAVALRY

19 GUN BATTERY

GRENADIERS

HUKELOM REGT

GRENADIERS

HANOVERIANS

PRINCE
OF PRUSSIA

CAVALRY

CONVOY

CASTLE OF
WYNENDAEL

REAR GUARD

Marlborough, who only had sufficient ammunition at Lille to last 4 more days. 'If our convoy had been lost', he told Godolphin, 'the consequence must have been the raising of the siege.' Sir John Fortescue wrote: 'The signal incapacity displayed by the French Commander did not lessen the credit of Webb, and Wynendael was reckoned one of the most brilliant little affairs of the whole war'.

The Allied force, without artillery and with only a single squadron of cavalry, inflicted a resounding defeat upon a vastly superior French force because of its great superiority in discipline and tactics. The Allied fire was incomparably more deadly than that of the French, partly because the Allies had a better musket, but more because they fired by platoons whereas the French fired by ranks. With a maximum range of about 300 yd, the flintlock musket was so inaccurate that eighteenth-century European soldiers trained for volley-firing at 60–80 yd; in close formation they were able to hold off an attacking force by the sheer volume of their two aimed rounds per minute. The drilled and disciplined manner in which the Allies' superior musket was handled was superbly manifested in this overwhelming victory at Wynendael.

The Allied victory at Oudenarde, just over 2 months earlier, had given an enormous boost to Allied morale, and the French were discomfited by the aura that surrounded Marlborough's name. To both sides he was indisputably the greatest commander of his day.

Rating of Commanders and Observations
Of the commanders in the field at Wynendael, Generals Cadogan, Webb and Lottum are 'above average', while de la Mothe should be rated 'below average'.

The events leading up to the battle can be marked on a scale-map covering an area of about 12 miles from north to south, with Roulers in the south, Ichtegem and Wynendael in the north and Thourout in the middle. The normal marching speed of infantry of this period was about 10 miles per day (cavalry about double that), although Webb's infantry could march to the battle area at 'forced' march rate of 15 miles per day, penalised by an agreed percentage (perhaps 10 per cent or less) dropping out en route.

The Wynendael position bore a strong resemblance to Agincourt, where in 1415 Henry V, with a force the same size as Webb's, overwhelmingly defeated a French army about as strong as de la Mothe's. Webb improved upon Henry V's dispositions by placing concealed troops so as to take the on-coming French in flank. The Allies were in an unflankable position possessing all the cohesive merits of the defensive square, plus the ability to bring maximum firepower to bear. The French cavalry charge at Wynendael was as unsuccessful as the charge of the French knights at Agincourt.

The cannonade lasts for a period in the wargamers' time-scale equivalent to 2 hours, and allowances must be made for Webb's tactic of ordering his troops to lie down when assessing casualties. The first line of the French infantry attack might have been composed of 1 regiment, other regiments subsequently following in line, or the 34 French infantry battalions may have attacked in battalion or even regimental columns—important factors from a wargaming point of view. On receiving fire, the French will check their state of morale, and then continue to advance or retreat, perhaps in disorder. If the entire front line is formed of 1 battalion and its morale is bad, that complete line will retire, with detrimental effects upon the lines behind it. However, if the attack is made in column, 1 battalion-column may be forced to withdraw but there will be others whose state of morale will be good enough to permit them to advance.

As at Agincourt, the restricted terrain prevents the French commander from bringing all his forces into action simultaneously; his formations are compressed and crowded upon each other in any case as they funnel into the wooded defile. They also then come within range of the troops posted in the woods on either flank. There is nothing harder to portray on the table-top than concealment, and the French 'commander' is well aware that there are troops concealed on his flanks; but, for an accurate simulation of the Wynendael skirmish, he must advance as though he were unaware of that fact. Such a course, however, will almost certainly lead to defeat; so a compromise is required.

A rudimentary simulation of an ambush on the wargames table is to allot one plastic counter of the same basic colour as the uniform of the man concerned for each figure involved in the ambush. Webb conceals say a dozen red-coated British infantrymen in the woods on his right flank, represented by that number of red counters similarly concealed in the table-top forest. The French may, when within 6 in of a visible counter, deviate or take some form of offensive action, and it has to be decided whether the ambusher or the attacker fires first. In their excitement the ambushers may fire too soon, or they may be steady enough to hold their fire until the oncoming French are at point-blank range. When the ambushers fire into them, the packed French ranks recoil towards the centre, so that all units affected by this flank fire must test their morale.

The French Dragoons rallying and supporting the infantry will raise their morale, and they may join in a mêlée with the Allied infantry formations. In the table-top reconstruction the result of this mêlée may turn in favour of the French, so that they force back the entire British line rather than breaking through the first line of 2 battalions but being forced back as the gap is filled from the second line, as occurred in 1708. The French were, in fact, thrown back in some confusion, and de la Mothe sent forward 8 more lines of infantry, who, in a wargame, would need to test their morale—undoubtedly they would have been shaken by the repulse of their comrades. When Webb's Allied infantry drove them back in panic, the French infantry's morale had reached its lowest ebb.

When the French are fleeing in disorder, Webb orders his infantry to advance, and they move forward by platoons, each firing in support of the other. A Military Possibility might disagree here. Did the Allied infantry have sufficient ammunition left to continue firing or had they expended it all? Infantry carried 23–30 ready-prepared cartridges, and in a lengthy action commanders had to make sure that their troops did not fire off their ammunition too quickly. This factor can be simulated in a wargame by allocating, say, 8 rounds of ammunition per man at the start of the battle and then noting the number of volleys, so that, as in real-life, the careful commander must exercise fire control.

The last phase of the battle brings Cadogan with what Fortescue calls 'a few squadrons of cavalry' but, probably because of the uncommitted French squadrons hovering around at the rear of the battlefield, no pursuit is pressed home. This presupposes that our table-top simulation has followed the same course as the original battle—but what if the boot is on the other foot and the Allies are in dire straits? Now Cadogan's cavalry might be needed to aid a sorely pressed Webb, and it would be necessary to define exactly how many were Cadogan's 'few' squadrons. They could move on the map, and arrive on the wargames table at the right moment; Chance Cards could be used to determine if any of them fall by the wayside.

Other variations are open to the French 'commander' who, with hindsight, will not wish to follow the losing tactics of Count de la Mothe. For example, French cavalry could move on the map to intercept the convoy on its circuitous route away from the scene of action; or Webb could weaken his force at Wynendael by detaching infantry to guard the convoy—there is little information as to the escort but obviously the convoy was not unguarded. Both these Military Possibilities, however, would turn the battle into just another wargame.

Construction of Terrain

The bulk of the battle took place in the defile between the woods, which sheltered Webb's flanking regiments. As a Military Possibility might cause the French to go in after them, the woods should be constructed to allow for this eventuality. They should be represented by large sheets of irregularly cut hardboard, painted dark green, their edges irregularly lined with

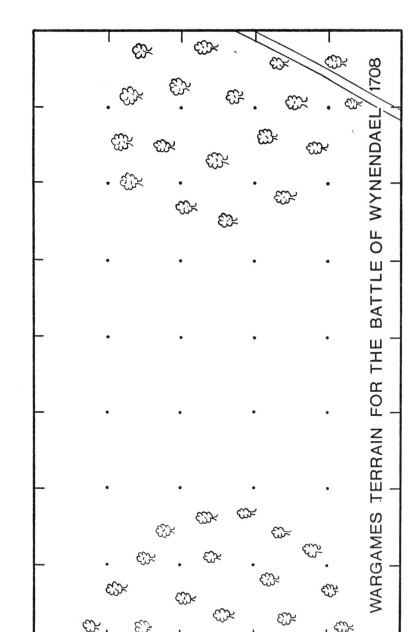

WARGAMES TERRAIN FOR THE BATTLE OF WYNENDAEL 1708

trees. On the bottom right-hand corner of the terrain you could mark in the road over which Webb reroutes his convoy. This road can be shown and possible French attempts to intercept the convoy can be made on the map.

6
The Battle of Prestonpans
21 September 1745

THIS ENGAGEMENT was fought during the Jacobite rebellion between the Highland army of Charles Stuart, the Young Pretender, and an English force commanded by Sir John Cope who, following the fall of Edinburgh, marched hurriedly from Dunbar to bring the Jacobite army to action. On 20 September 1745 the two armies found themselves about $\frac{3}{4}$ mile apart and, aware of the Highlanders' headlong methods of attack, Cope took up a defensive position at the eastern end of a flat and featureless tract of land lying to the north of the higher ground around Tranent.

Prince Charles' force numbered about 2,550 infantry and 40 mounted men, but lacked artillery. The Highlanders were armed with a broadsword, a dirk and a target (a small circular shield); some had pistols and a few had muskets. Their strength lay in their frightening headlong charge. Cope had about 3,000 ill trained and ill disciplined troops, including 6 squadrons of dragoons, whose horses were unsuitable for their role, being unaccustomed to firearms and frequently taking fright and bolting. His lack of Regular artillerymen greatly handicapped Cope; although the Highlanders had become accustomed to artillery, its value was more than demonstrated by Cumberland at Culloden, and its effective use would have heightened morale at Prestonpans, where the English had six $1\frac{1}{2}$ pdr guns and 6 mortars. The redcoats had muskets with bayonets, but their slowness in reloading brought disaster at Prestonpans, though the lesson was learnt in time for Culloden, where the

English infantrymen, formed in three ranks and firing one rank at a time, effectively thrust their bayonets in the unguarded sides of the Highlanders.

Anticipating an attack from the south, Cope positioned his men with his right flank towards the walls of Preston House. His artillery was posted all together on the left wing instead of being distributed along his line, since he only had some sea-men-gunners from warships, who were frequently drunk, to man the guns, and they ran away at the start of the action.

The Highland army prepared for a fierce charge, but close examination of the ground revealed that any such attack was quite impracticable. Even at this stage there was disagreement between the army commanders, Lord George Murray and John O'Sullivan; there was mutual distrust, and much doubt from O'Sullivan when Lord George outlined a bold but simple plan for a night march round the east end of the morass to fall upon the enemy's open flank, after a local man had described a track that would bring them to the area.

In his initial position Cope attempted to shake the High-landers with artillery fire, running two of his $1\frac{1}{2}$ pdrs forward to be loaded and fired by the seamen-gunners, but at a range of 800 yd the little roundshot, with unreliable fuses, were ineffec-tive. Similarly, during the night, Cope attempted to annoy the Highlanders by shelling their position, but abandoned the idea after a single mortar round revealed the inadequacies of the fuses.

Commanding a numerically superior force of Regular soldiers, Cope might be criticised for not attacking the High-landers, though this would have exposed his army to the same disadvantages as would have accompanied an attack by the Highlanders in the opposite direction. None of Cope's troops had seen previous action, and they could only be kept in hand in a defensive position; the ground was suited for cavalry and the Highlanders had none.

Shortly before 4 am, led by the local guide, the Highland army

E

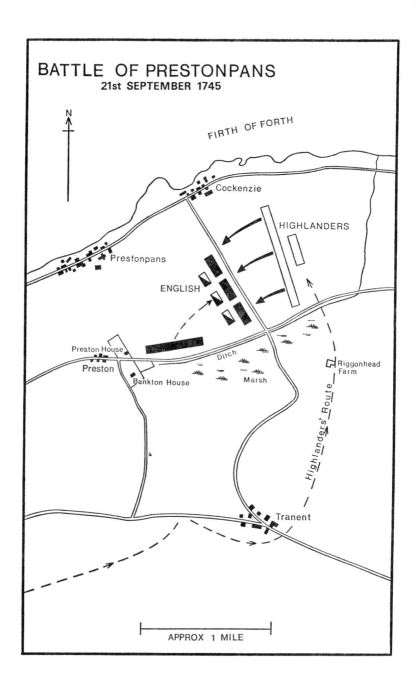

began to move slowly north-eastwards down the ridge; the column passed safely through the narrow defile close to the farm of Riggonhead, where it could have been intercepted by Cope's cavalry scouts. Dawn came as it traversed the morass, and despite the low morning mist it was discovered by patrolling Dragoons, who galloped back to raise the alarm. Passing over the narrow plank bridge that crossed the 4 ft ditch bounding the morass on the north, the Highland vanguard debouched on to the plain about 1,000 yd to the east of Cope's left flank. The head of the column marched on towards Cockenzie until it was estimated that the rear of the front line troops were clear of the marsh but, when the van turned left to form line it was separated from the rear by a gap that was not closed until just before the attack began.

Receiving the dragoons' report of the Highlanders' movement, Cope had an alarm gun fired to recall his patrols, and then manoeuvred his army to face the enemy, with the infantry wheeling left by platoons and marching off northwards, roughly parallel to the line of march of the Highlanders. Halting, the infantry turned right into lines three deep and faced east, with their left towards Cockenzie and their right protected by the large ditch. Drawn up from left to right were 9 companies of Murray's, 8 of Lascelles', 2 of Guise's and 5 of Lee's; they were separated from the artillery on their right by a space sufficient for 2 squadrons of cavalry. With mortars on their right, the guns, 6 ft apart, were in line with the infantry, and were guarded by 100 men of Murray's Regiment on the right of the line. Two deep, a squadron each of Gardiner's and Hamilton's Dragoons, with one of each in reserve, moved up to take position in the line; Hamilton's formed up as ordered but Gardiner's formed up in rear of the artillery, where there was little room, for this space had been taken up by the infantry out-guards, who had returned but did not have time to rejoin their regiments.

The Highlanders' front line began to advance and Cope, seeing that his left wing was outflanked, ordered two guns to be sent over from the right, but the frightened civilian drivers rode

off with the limber horses, closely followed by the seamen-
gunners who were supposed to work the guns.

As the sun rose, Lord George Murray ordered the Highlanders of the left wing to advance; they moved forward un-
evenly, their left well in advance of their right wing. Discarding
their plaids, and with bonnets pulled low over their brows,
they rushed forward screaming, terrifying the makeshift artillerymen, who fled. The two officers left to handle the guns
managed to fire 5 of the 1½-pdrs and all 6 of the mortars,
although many of the defective shells burst ineffectively.
Had the guns been manned by Regular artillerymen their
effect might have been decisive, because even this
spasmodic fire caused a momentary tremor in the Highland
ranks.

The Highlanders broke into separate bodies, the largest
group making straight for the artillery. Hamilton's Dragoons
were ordered to charge them in the flank and their commander,
Colonel Whitney, led them out to within pistolshot of the
Highlanders, but suddenly all the horsemen turned and ran,
leaving Whitney alone. The artillery guard left their ground and
crowded in a confused mass behind the guns they were supposed
to be guarding and, when instructed to support the Dragoons,
refused to go forward. Quickly their officers reorganised them
into a regular front rank but, after delivering an ineffectual
volley, they gave way as both they and the artillery were overrun. Gardiner was ordered to charge but, alarmed by the
artillery guards falling back, his entire squadron turned and
fled.

On the extreme right the advancing Highlanders maintained
their line in very orderly fashion and, outflanking Cope's left,
swung obliquely inwards, and with their musket fire caused
casualties in Hamilton's other squadron, which galloped off in a
panic-stricken body.

Deserted by the cavalry and exposed to the full fury of the
Highlanders' attack, the unfortunate infantry wavered, in spite
of the efforts of Cope and other officers to steady it. The foot
soldiers were broken into from right to left by successive waves

of Highlanders, hacking and stabbing their way forward, and became a flying rabble. Cope and the Lords Loudon and Home tried to round up the dragoons, who were now crowded into the narrow defiles south of Preston House, and succeeded in turning about 450 of them into a field and reforming them. But then a body of Highlanders appeared and the cavalrymen refused to attack; so, realising that nothing could be expected of their men, Cope and the rest of his officers rode off at their head, this being the only way of keeping them together.

Scarcely more than 5 minutes had elapsed from the first onslaught on Cope's army to its rout. Cope's infantrymen, in their tight clothing and heavy equipment, had little chance of escape, and all but 170 of them were killed or captured. In all, Cope lost 200 killed and about 500 wounded, and 1,500 of his men were taken prisoner. Highland casualties were 30 killed and 75 wounded.

Although Prestonpans was a decisive Jacobite victory, it played a vital part in their defeat and destruction at Culloden by persuading Charles Stuart that his Highlanders were invincible.

Rating of Commanders and Observations
Cope was 'below average', O'Sullivan 'average' and Lord George Murray 'above average'. The British unit commanders (Murray, Lascelles, Guise, Lee, Gardiner and Hamilton) were all 'below average'.

Making Lord George Murray an 'above average' commander balances the numerical deficiency of the Highlanders, besides ensuring that their morale is sufficiently high to undertake the vital outflanking movement. A Military Possibility can represent the known discord between Lord George Murray and John O'Sullivan by the occasional drawing of Chance Cards to control the cohesiveness of the Highland army.

The key to the battle is the race between the Highlanders to deploy and charge while Cope, perhaps after a momentary check through surprise, turns his army left and marches it to face its new position. This march must be done in a reasonably formal fashion, as these were the days of unthinking soldiers moving *en masse* in response to rigid orders. In a sense the situation is akin to a Colonial battle in which fleet-footed natives outmanoeuvre their slower opponents and then, backed by the superior morale given them by their frightening weapons and courage, charge overwhelmingly upon their more civilised enemy.

If Cope's cavalry scouts had discovered the advancing Highlanders in the Riggonhead defile and fought a minor delaying action, the English commander could have had more time to reposition his army. This is a Military Possibility that could make a difference to the battle with only small amendment to reality. A delay advantageous to Cope could be caused by the Highland army funnelling in to cross the plank bridge over the 4 ft ditch.

The inadequacies of Cope's artillery, both in gunners and defective fuses, should be reflected by local rules giving a low standard of morale to the gunners and little effectiveness to the missiles. Cope's artillerymen fled at the sight of the enemy—a Military Possibility could give them a slight chance of staying and firing their guns. Assuming that they follow historical precedence and run, it must be decided how effective was the fire of the two officers who remained. It has been noted that even these despairing efforts caused a momentary tremor in the Highlanders' ranks so that sustained artillery fire might well have reduced their morale and checked their charge.

Cope's infantry out-guards, 300 strong, were made up of detachments from each regiment and, as we have said, did not have time to rejoin their own units, crowding out Gardiner's Dragoons instead. It is feasible to assume that, lacking the esprit-de-corps of being with their own units, this force had a low standard of morale.

As in real life, attacks on the wargames table may be held off by weight of firepower. The morale of the firers affects their aim and their ability to remain steady and hold their fire until the enemy are close upon them. Shaky, ill disciplined and badly trained troops will fire off a volley at long range, and then frantically reload their muskets so as to give the onrushing enemy another volley before they make contact. Vital aspects of this battle could be the inability of Cope's infantry to hold its fire, or the possibility of its reloading to get in two volleys. There are a number of Military Possibilities here worthy of exploration.

Well handled cavalry could be extremely effective against the Highlanders, who did not possess any horsemen, but Cope's cavalrymen, with their low morale, refused to attack. A Military Possibility could give them a higher rating, so that some of them will charge, which could affect the outcome of the battle. Cope's inability to rally his infantry could be changed by a Military Possibility that gave him a slight chance of doing so; or he might induce his reformed cavalry to charge from the area of the Preston House—a determined cavalry charge on the disordered and uncohesive Highlanders from there could have swung the battle in his favour.

This action was largely decided by morale—low among the English soldiers and high among the Highlanders. The former were afraid of the Highlanders and their fearsome onslaughts with the claymore. These charges can only be turned back on the wargames table by firepower *before* they make contact; once they force a mêlée their impetus (which can be represented by assuming that two Highlanders are the equal of three British infantrymen) will ensure success.

Construction of Terrain
This terrain must allow for Cope to take up his original position along the road running from Preston House to the crossroads, for the Highlanders to come on to the table in the later stages of their outflanking march, and for Cope to take up a new position facing his flank. There must be sufficient room for the action that follows to take place in the top right-hand third of the terrain, in an area bordered by the roads to Cockenzie and the ditch. It is not essential to have the buildings on the table, but they provide areas which might be affected by Military Possibilities, besides giving interest to the appearance of the

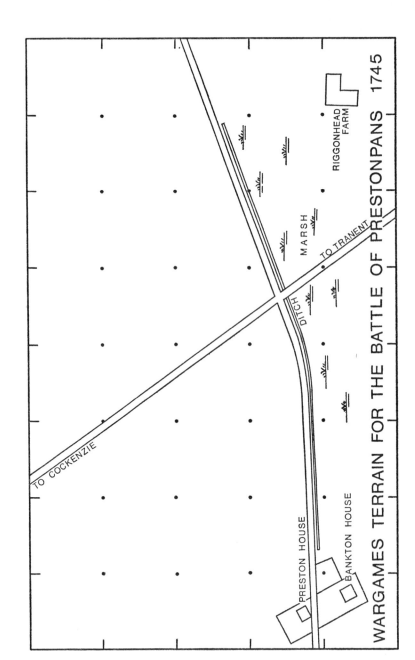

WARGAMES TERRAIN FOR THE BATTLE OF PRESTONPANS 1745

RIGGONHEAD FARM

MARSH

TO TRANENT

DITCH

TO COCKENZIE

PRESTON HOUSE

BANKTON HOUSE

terrain. The ditch and marsh, while not playing a vital part in the historical conflict, could, with some Military Possibilities, take on some importance. This is not a difficult terrain to make, as it is completely flat.

7
The Battle of Guilford Courthouse
15 March 1781

THIS ENCOUNTER took place during the American Revolution (1775-83) between an American force under General Nathaniel Greene and a British army under General Cornwallis. The contrast between the rigidity of the British Regular forces and the loose and unorthodox tactics of the American Militia makes for unusual but fascinating table-top battling.

This British victory marked the beginning of the end of the Revolution, because, by winning, Cornwallis so weakened his army that he lost the campaign in the southern colonies. Cornwallis had been pursuing Greene in the Carolinas in the hope of forcing an engagement, while Greene, wishing to avoid battle, was trying to draw the British as far as possible from their base. Early in March Greene collected reinforcements in Virginia that strengthened his force to about 5,500 men, moved westwards to Guilford Courthouse, within 12 miles of the British Army, and took up battle stations facing west in three lines 400 yd apart. At the head of a force of 1,900 men with 3 guns, Cornwallis marched at daybreak on 15 March. About 4 miles from Guilford Courthouse the advance cavalry of both armies met in a brief skirmish, and the Americans were driven back. Continuing the advance along the New Garden Road, the British crossed a small stream from which the ground rose gradually to an open space about 500 yd square, made up of three cultivated fields bordered by rail fences; the whole area forming a defile between thick copses of trees. Here, with an excellent field of fire and protected by the rail fences, Greene had drawn up his first line of 1,600 men. This line consisted of two brigades of North Carolina Militia, untrained soldiers without battle experience; two battle-experienced units — Lee's Legion (Regulars) and Campbell's Riflemen (frontiersmen from Virginia and the North Carolina mountains)—on their left; and Colonel William Washington's Regular Cavalry, some of the Delaware Regiment of Continentals and Lynch's Riflemen (veteran frontiersmen) on their right. Two 6-pdr guns in the centre on the road commanded the stream crossing.

About $\frac{1}{2}$ mile ahead of the clearing, in the woods that closed in on the road, Greene deployed his second line of about 1,000 Virginia Militia, untrained and inexperienced. Commanded by General Wilson on the right and General Stevens on the left, and officered by men who had served in the Continental army and had some battle experience, this second line was somewhat stronger than the first. All the troops of the second line were hidden in the woods on each side of the road, with picked marksmen behind them, ready to shoot any man who ran away.

Then came another open space of cultivated ground, made uneven by hollows, and 400 yd in rear of the second line, and rather to the west of the road, was a hill that formed the salient angle in the midst of the clearing around the Courthouse. Near the eastern edge of this clearing Greene drew up his third line of 1,450 Regular troops, the 2nd Virginia Brigade on the right and the 2nd Maryland Brigade on the left, with two 6-pdr guns between them. The right flank was unprotected, but the left flank, resting on the New Garden Road, was protected by artillery during the later stages of the battle. Both flanks of the first two lines were unprotected, but as the heavy woods forced the British to attack frontally, these exposed flanks were not a disadvantage.

Cornwallis' army was much better organised, disciplined and trained than the American, and his men, perhaps the best British forces in America, were veterans commanded by able ex-perienced officers. Their right wing under Major-General Leslie consisted of Bose's Hessian Regiment and Fraser's Highlanders in the first line, with the 1st Guards in support; the left wing

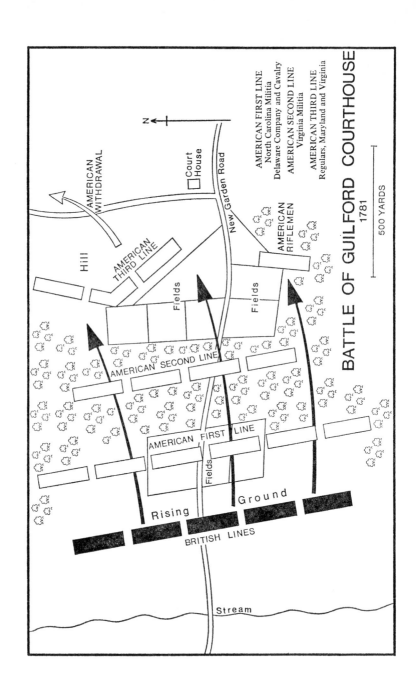

BATTLE OF GUILFORD COURTHOUSE
1781

AMERICAN FIRST LINE
North Carolina Militia
Delaware Company and Cavalry
AMERICAN SECOND LINE
Virginia Militia
AMERICAN THIRD LINE
Regulars, Maryland and Virginia

500 YARDS

AMERICAN WITHDRAWAL

Court House

New Garden Road

AMERICAN RIFLEMEN

Hill

AMERICAN THIRD LINE

Fields

Fields

AMERICAN SECOND LINE

AMERICAN FIRST LINE

Fields

Rising Ground

BRITISH LINES

Stream

N

was formed of the 23rd and 33rd Regiments under Colonel Webster in the first line, with the Grenadiers and the 2nd Guards in support. A small corps of German Jagers and the Light Infantry were stationed in the wood to the left of the road, while Tarleton with the cavalry remained in the rear on the road. Cornwallis posted his 3 guns on the road itself; they could not move anywhere else because the woods restricted their field of fire.

The British force formed up under American artillery fire after crossing the stream at the foot of the hill, losing very few men, while the British artillery replied with an equally useless expenditure of ammunition. At about 1.30 pm the British troops advanced across the first clearing, amid a hail of fire from the invisible enemy in front and on the flanks, then charged with the bayonet. The North Carolina Militia melted away in panic, while Greene's 2 guns retired to take post near the road with the American third line. The British line halted and the flank battalions turned outwards to oppose the American riflemen, who were still pouring in a destructive fire from both sides. Bose's regiment wheeled off to the right flank, and the 1st Guards moved into the line to take their place next to the Highlanders; on the other wing the 33rd, with the Jagers and Light Infantry on their left, similarly wheeled off to the left flank and the Grenadiers and the 2nd Guards advanced to the left of the 23rd.

Slowly the redcoats pressed forward through the trees, driving the American riflemen back at the bayonet point; the cavalry of Lee and Washington fell back to conform. Bose's Regiment and the 1st Guards drove Lee and Campbell's left flank detachment backwards in a south-easterly direction in a struggle that completely detached them from the main course of the battle, their separate engagement only being broken off by the Americans at about the time the main conflict ended. The American right flank detachment briefly took position on the flank of the second line, and, when that retired, moved to the flank of the third line.

The attack proceeded east along the road and through the

woods for about 400 yd until it struck the second line, where the rifle-armed Virginia Militia stubbornly held their ground, inflicting heavy losses on the advancing redcoats. Discipline and experience began to tell, and gradually Wilson's Brigade on the American right crumbled before the Light Infantry and Jagers, who were their equals at bush fighting. Stevens' Brigade, on Wilson's left, stood firm for a while, but finally the pressure told and both forces were pushed back until, reaching the road, they broke and retired in disorder through the wood to Greene's third line, protected from pursuit by Washington's cavalry retiring with them to the edge of the forest.

Entirely north of the road, Greene's last line was opposed by the British left wing, which was slowed up by heavy undergrowth and deep gullies. Led by Webster, the now considerably reduced 33rd, Light Infantry and Jagers, attacked the 1st Maryland Regiment, the finest battalion in the American Army, which steadily awaited the assault before pouring in a volley at close range and charging with the bayonet to drive the attackers back in disorder, with heavy loss. Although severely wounded, Webster drew his men off into the shelter of the woods and rallied them, supported by the 3 British 3-pdr guns, which advanced along the road and unlimbered on the rising ground at the edge of the forest, whence they kept up a steady and well directed fire.

While Webster's force was rallying, General O'Hara led the 2nd Guards and the Grenadiers, supported by the 23rd and the Highlanders, against the Maryland Brigade. Seemingly giving way before O'Hara's attack, the 2nd Maryland Regiment steadily withdrew as the Guards pressed them; suddenly Washington's cavalry galloped out of the woods to crash into the rear of the Guards, while the 1st Maryland Regiment flooded out of the undergrowth on to their left flank. Fighting fiercely, the Guards were utterly broken, and the danger of a wholesale retreat forced Cornwallis to order his artillery to fire grapeshot into the struggling mass, causing great losses to both Americans and British.

In the pause that followed Cornwallis reformed his line,

rallying the 2nd Guards, who had been joined by the 1st Battalion; Webster led up the 33rd, the Jagers and the Light Infantry on their left. Under the smoke of a volley Tarleton's Cavalry were sent in, to end the American resistance on the British right and then they moved across to join Bose's Regiment and press upon the Americans' left flank, while the remainder of the British line engaged their front.

Seeing that the day was lost, Greene ordered retreat, abandoning his guns; there was no pursuit by Cornwallis' exhausted troops. Casualties were 93 British killed and 439 wounded; 78 Americans killed and 183 wounded.

Rating of Commanders and Observations

Greene undoubtedly displayed strategical sense by fighting not to win but to cause Cornwallis such losses as to diminish his future prospects of victory. Greene is certainly an 'average' commander but could be classified as 'above average' here, though that rating would probably put victory completely beyond the reach of Cornwallis (himself an 'average' commander), with his considerably smaller army. Webster could be 'above average', as could Washington; and other commanders can be classified in relation to the performance of their units. It will be necessary to assess the varying states of morale and fighting ability of

(a) the British Regulars,
(b) the Continental Regulars,
(c) the American Riflemen, and
(d) the American Militia.

This assessment could be handled after the manner of the Wargames Research Council's Ancient Rules, where Regulars, Barbarians and levies, etc, are given different values when building up an army. Otherwise, the British Regulars and the Continental Regulars could be considered as top-grade soldiers; the Riflemen classed as light infantry, but not given the high morale of the Regulars, because of lacking the Regulars' cohesion; and the militia given a low grading, both in morale and fighting ability, because of the likelihood of their fleeing the field without firing a shot.

With the numerically superior American army in positions of their own choosing and the British historically pledged to attack in rigid formations, the only manner in which this battle can be accurately simulated is to devise 'local' rules that balance these factors. The Militia must check their morale state as soon as the British come within charge-move distance, whenever they take fire and sustain casualties, or whenever they are brought into mêlée contact. The British Light Infantry and the Jagers should be given the same powers of manoeuvrability as the American Riflemen.

The cohesion of the British units attacking in rigid formations should give them an increased morale bonus, or else it is doubtful whether they will ever get into mêlée contact, as they did in real life. In fact, the morale of the British should be capable of being raised to great heights; particularly in the case of the Guards, who were driven off after being severely mauled by the high-grade Maryland Regulars, hit in the rear by cavalry and then flailed with grapeshot by their own artillery. They subsequently rallied and came forward again to the attack—a rare exploit, unlikely to occur under normal morale rules.

Military Possibilities can be devised to prevent the Militia from fleeing, to prevent the American second line from being pushed back, to alter the withdrawal of Webster's men and even their subsequent rallying, and to allow the Guards to beat the Maryland Regiments and to turn and fight off the cavalry rear attack, etc, etc. But the historical reality of Greene's de-laying action to maul his opponent, who consistently pushed forward, indicates that the most realistic simulation will be achieved by judicious handling of the morale situation.

Construction of Terrain

This battle, being fought in three separate stages, requires the extensive area allowed by laying the terrain lengthways, with the stream on the 'bottom' baseline. From there a ridge, 18 in wide on its top, stretches right across the table, with 3 rail-

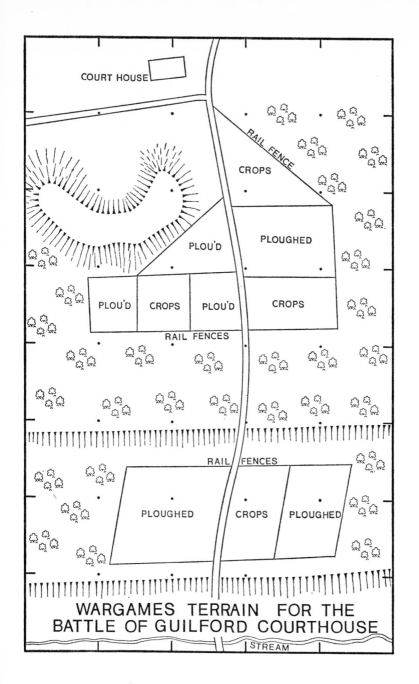

COURT HOUSE

RAIL FENCE

CROPS

PLOU'D

PLOUGHED

PLOU'D CROPS PLOU'D CROPS

RAIL FENCES

RAIL FENCES

PLOUGHED CROPS PLOUGHED

RAIL FENCES

WARGAMES TERRAIN FOR THE
BATTLE OF GUILFORD COURTHOUSE

STREAM

fenced fields, each about 18 in square. Because fighting has to take place amid the trees, the comparatively spacious areas of woodland must be made 'symbolic', by representing them with hardboard bases bearing occasional trees around their edges, the whole being the wood itself. The extensive ridge and hill are easy to form by draping grass-coloured cloth over shapes below. Considerable lengths of rail fence are required; this can be made from basket-weaving cane.

8
The Battle of Maida
4 July 1806

THIS BATTLE was fought between a British force under General Sir John Stuart and a French army commanded by General Jean Reynier in Calabria during the Napoleonic Wars, undoubtedly the favourite period of wargamers, though most of the battles are too large for the wargames table. Maida, however, has all the typical features of the time.

On 1 July 1806 a British force under Sir John Stuart landed at the Bay of St Euphemia. It consisted of 7 battalions of infantry organised into an advance corps and 3 brigades; 2 of the battalions were composed of flank companies of various regiments and were experienced veterans, as were the 20th and 27th battalions, but the three remaining battalions comprised raw recruits. Totalling about 5,500 men, the force had eleven 4-pdr guns but no cavalry. The 20th battalion was detached to make a feint attack on Reggio.

Instead of gaining a tactical advantage by falling upon Reynier's troops while they were still dispersed, the British general remained stationary throughout 1, 2 and 3 July, because of the heavy surf, which made the landing of provisions difficult. Reynier gathered his force together during the 3 days, marching 80 miles from Reggio to Maida, where he arrived on 3 July at the head of a force totalling 6,400, formed of 6 French infantry battalions (4,123), 1 Swiss infantry battalion (630), 2 Polish infantry battalions (937), 328 cavalrymen and a battery of horse artillery.

On the morning of 4 July the British force marched on to the plain of Maida in echelon of brigades with Kempt's Light Infantry leading the way, its right flank skirting the thicket that bordered the Lamato River; next came Acland's Brigade and then Cole's on the extreme left, while Oswald formed the reserve in rear of centre with 12 companies of infantry and 3 field guns. Manoeuvring in their front the French cavalry and horse artillery raised much dust, and the smoke of their guns and the British field pieces exchanging shots obscured the movements of the French infantry. When the dust subsided, Compére's Brigade of 2,800 veterans of the First Light Infantry and the 42nd of the Line advanced rapidly to the attack. On Compére's right was Peyri's Brigade of 1 Swiss and 2 Polish battalions (1,500) while on the right was Digonet's Brigade which included 1,250 infantrymen of the 23rd Light besides the cavalry and artillery.

Formed in line, the British 2 deep and the French 3 deep, the armies approached each other under a spasmodic artillery fire, the French, unlike the British, gunners causing few casualties. When Kempt's and Compére's infantry were less than 100 yd apart, the leading French battalion halted and fired a volley; both lines exchanged some 2 or 3 rounds before Kempt moved forward again, only to halt for his men to remove their greatcoats. The French mistook the movement for the beginning of a retreat and started to rush forward; Kempt sent his men slowly onwards, and when the French were within 30 yd, ordered: 'Fire! Charge bayonets!' Before the British could charge, however, the foremost French battalion turned and fled, taking away with them the rear battalion of the First Light; then the entire brigade turned and ran, pursued by the British. The pursuit lasted for more than a mile until, at the village of Maida, Kempt succeeded in halting and rallying his men. The French First Light had sustained 900 casualties, Kempt less than 50, but the British Light Infantry Brigade had put themselves out of the action by their pursuit.

Acland's Brigade took a volley from the 42nd French opposite them, and then delivered a destructive counter-volley, and the four British 4-pdrs upon the brigade's right poured in

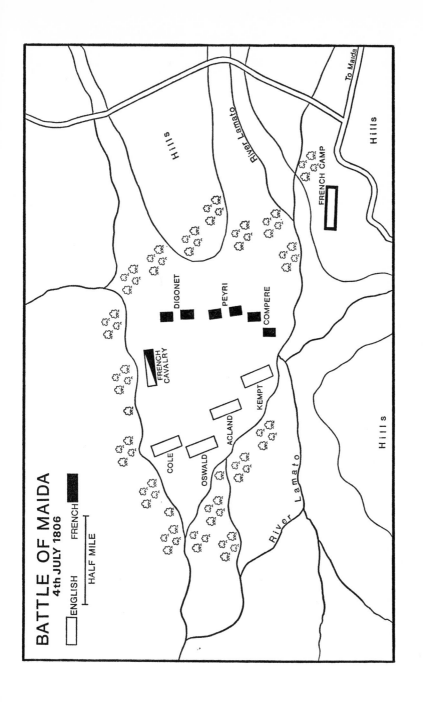

BATTLE OF MAIDA
4th JULY 1806

ENGLISH ☐   FRENCH ■

HALF MILE

DIGONET   PEYRI   COMPERE

FRENCH CAVALRY

KEMPT

ACLAND

OSWALD

COLE

River Lamato

River Lamato

To Maida

FRENCH CAMP

Hills

Hills

Hills

Hills

grapeshot. The French regiment gave way and ran, but later rallied and took up a position near Reynier's right wing. The French commander now brought up Peyri's Brigade, but the Poles gave way at once before the 81st. The Swiss, wearing red coats, were mistaken by the Highlanders for British, and were allowed to approach and pour in a damaging close range volley; the 78th rallied and their volleys of musketry drove the Swiss back. With the Highlanders in the lead, Acland's Brigade moved forward, harassed by demonstrations from mounted chasseurs and horse artillery.

In the meantime Cole's Brigade was coming up on the extreme British left, having been much delayed by French cavalry and Horse Artillery, which had compelled them more than once to form square. Reynier formed the 23rd upon rising ground and threatened a cavalry charge, while detaching his Light Companies to harass the British battalions from the brushwood on their left flank. His ammunition running short and his men exhausted from their efforts under the hot sun, Cole was hard pressed, but was relieved by the arrival of Ross with the 20th Foot, which had landed and doubled up to the sound of firing. The men plunged into the brushwood on Cole's left, drove out the sharpshooters, poured a volley into the French squadrons to send them to the rear in confusion and then, wheeling to the right, opened a shattering fire upon the flank of Digonet's battalions. After a feeble attempt to hold his ground, Reynier drew off his troops and retired from the field, skilfully covering his retreat with his cavalry and sharpshooters.

Had the British possessed even 2 or 3 squadrons of cavalry, hardly a soldier in the French army would have escaped. Even so, Reynier's losses were severe—more than 2,000 killed, wounded and prisoners compared with Stuart's 45 killed and 282 wounded.

Maida was a soldier's battle, won by the sheer merit and fortitude of individual units and their commanders, plus the superiority of British musketry training, and owing nothing to the skill of their commander, whom Fortescue described as 'cantering about all over the field . . . enjoying himself as a spectator but giving not a thought to the direction of the battle'. Fortescue also remarked that any one of the 4 British brigadiers —Cole, Kempt, Oswald and Ross—would have made a better commander than Stuart, and all made their mark later in the Peninsula.

Rating of Commanders and Observations
Reynier, the French commander, does not show up at all well, and could be fairly given a 'below average' classification. His British opponent, Sir John Stuart, cannot be given better than 'average'—which makes it difficult for the numerically smaller, cavalry-less British force to achieve its historical success. It might be best to make both commanders 'average', and to give the British and French regiments morale and fighting classifications based on their performances in the battle.

British success was due to high morale coupled with the fact that the British musket threw a heavier bullet than the French, giving it a greater disabling power. The British fought in double and the French in triple ranks, so that every British musket was brought into action, thus presenting a greater front of fire—a 600-strong British battalion occupied a front of 200 yd, whereas a battalion of 600 French soldiers covered only 135 yd, allowing the British to overlap the French by more than 30 yd each side and enabling them to fire on flanks as well as front. The British soldier had been carefully trained and strictly disciplined to obtain the maximum advantage from each shot. Both musket-balls and artillery fire could disable 3 French infantrymen, because of their formation, but the return fire could only disable 2 British.

All these factors must be reflected by suitably adapting rules and casualty-effect tables, together with the insistence that the British and French line formations are maintained throughout the battle. The British artillery would seem to have been better handled than the French, and a similar shading of the rules must reflect this.

Common to all horse-and-musket battles, the field soon became obscured by dust and smoke; at Maida the manoeuvring of the French cavalry and horse artillery caused heavy dust clouds to arise. This 'fog of war' is difficult to simulate, but there are helpful suggestions under 'Surprise' (pp. 20–21).

One vital aspect of the battle was the conflict between the British and the French Light Infantry, which, if it is to be realistically simulated, must have the French charging precipitately (handled by an uncontrolled advance, as in Wargames Research Council Rules, or by an indifferent morale rating or by Chance Cards). As they go forward, the French are (1) fired on by the steady British Light Infantry, and (2) charged by them. Morale checks will be necessary to study the effects of these two actions, or their cumulative affects could be handled by a single morale check. A Military Possibility can be devised giving the French Light Infantry a chance of holding firm, but if history is repeated and they break, with the British Light Infantry in pursuit, then it has to be decided just how far and under how much control the pursuit is carried on. In the original battle the British Light Infantry were taken out of the action by their pursuit, which can be simulated, or else, by means of morale checks or Chance Cards, they can be rallied earlier and return to take part in the remainder of the battle.

The low morale of some units on the French side, such as Peyri's Polish Brigade, must be reflected in local rules to balance up the numerical inequality of the forces. Also, no self-respecting wargamer commander is going to allow the red-coated Swiss troops fighting for the French to approach and fire a volley upon a British unit that mistook them for comrades. While it may not have a vital affect upon the battle's result, the situation should be simulated, and could be resolved by Chance Cards. The British unit mastered their surprise to rally and drive the Swiss back, indicating high morale or a high rating for their commander.

Both Acland's and Cole's Brigades were, at various stages of the battle, held in check by French demonstrations with mounted chasseurs and horse artillery. Undoubtedly, in accordance with the practice of the times, they formed square under the threat of the cavalry, and were then assailed with

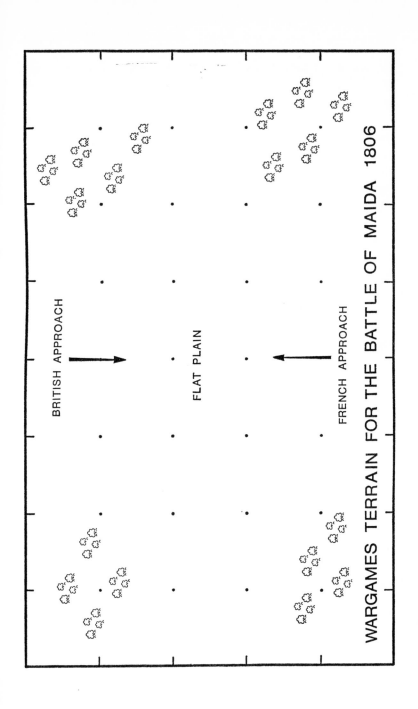

BRITISH APPROACH

FLAT PLAIN

FRENCH APPROACH

WARGAMES TERRAIN FOR THE BATTLE OF MAIDA 1806

grapeshot by the artillery. Nevertheless, the British brigades were not broken nor did the French cavalry attack them—both features are not difficult to simulate on the wargames table.

A vital aspect of the battle was the arrival of the 20th Foot on Cole's left, after a march that can be carried out on the map, possibly affected by Chance Cards, so that they will arrive in time or too late. This battle is highly suitable for some interesting pre-conflict map-manoeuvring, although, if that is not carefully controlled, it may affect the accuracy of the reconstruction.

Construction of Terrain

Maida has perhaps the simplest terrain of all the battles fought in this book, as it consists of a perfectly flat plain adorned by 4 areas of woodland. These are fairly extensive and, in at least one case, fighting or manoeuvring takes place within the confines of the trees. Thus, it is suggested that each area of woodland is 'symbolically' represented by an irregularly shaped piece of hardboard of the size indicated on the map, with a few trees dotted around its edges. The wood covers the entire area of the hardboard shape.

9

The Battle of Aliwal

28 January 1846

THIS WAS a brilliant action in which the powers of infantry, artillery and cavalry were successively and successfully brought into play, so that Sir John Fortescue, the historian of the British army, wrote '. . . it was the battle without a mistake'. As a Colonial horse-and-musket affair, this battle of the First Sikh War would be difficult to better.

The Honourable East India Company's force, commanded by Sir Harry Smith, consisted of 10,000 infantry (two-thirds natives), 2,000 cavalry and 32 guns. The Sikh force, led by Ranjur Singh, had 18,000 infantry, 2,000 cavalry and 67 guns. The battlefield was a level grassy plain some 2 miles long and 1 mile wide, with the River Sutlej in the rear; at the edge of the plain was a gentle rise, between Bhundri on the Sikh right and the mud village of Aliwal, and the two villages were connected by waist-high earthworks, curving along the ridge and masked by a thin grove of trees. Holding both the villages, the Sikhs were positioned on the crest of the rising ground, their guns spaced along the front of their line, facing south-east; between the river and the ridge lay the tents of their encampment.

At 10 am the British force began to move forward in an order of battle reminiscent of the Napoleonic Wars, each unit in line of column, the cavalry on the flanks and the artillery in the intervals. Commanded by veterans of the Peninsular War, the force's arms and equipment had not changed since Waterloo and, because of the northern Indian winter climate, the army wore the old-style European dress.

Sir Harry Smith intended to carry Aliwal first and then continue the attack upon the Sikh left and centre, cutting their line of retreat by the river fords. Under heavy artillery fire, the brigades of Godby and Hicks went forward towards Aliwal, taking a ragged volley at 150 yd from the Irregular hillmen manning the position, who, lacking time to reload, were overwhelmed as the advancing infantry broke into the village, capturing 2 guns and killing all the gunners. Now a general attack was made upon the Sikh left and centre by Wheeler's Brigade, supported by Wilson; Wheeler advanced under heavy artillery fire, halting twice and lying down to steady the men and to allow Wilson's Brigade to draw level. When they were close to the Sikhs, the attackers fired a heavy volley, causing many enemy infantry to retreat, although the gunners continued firing their guns. Throughout, the Horse Artillery flashed in and out, pausing to fire, then limbering up and retreating out of range before being hit.

His left in serious danger of being turned, Ranjur Singh changed his front, pivoting on the village of Bhundri, and covering the movement with a large body of cavalry, who were immediately charged and routed by a Native Cavalry Regiment. Hicks' Infantry Brigade went forward to support the cavalry while Godby's Brigade hit the Sikh left flank and rear, carrying everything before them. With his retreat via the river fords seriously threatened, Ranjur Singh sent forward another large body of cavalry to cover him as he threw back his left and, using the village of Bhundri as a pivot, reformed his line at right-angles to the river.

A squadron of the 16th Lancers ordered forward to attack the Sikh horsemen sent them streaming away towards the river in full flight. Returning, this squadron broke through squares of crack Aieen Sikh infantry trained by French mercenary officers.

Wilson's Brigade went forward to attack an artillery battery still firing resolutely from the centre of the Sikh position, but took such heavy losses that the native battalions of the brigade began to waver. Brigadier Cureton ordered the 16th Lancers to

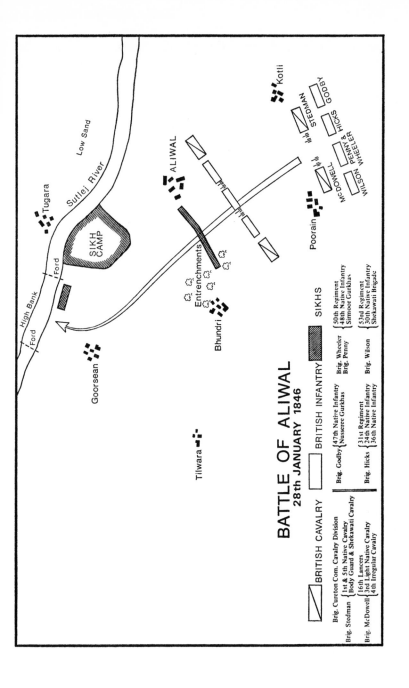

BATTLE OF ALIWAL
28th JANUARY 1846

Tugara

Low Sand

Sutlej River

High Bank

Ford

Ford

Ford

SIKH CAMP

Goorsean

Tilwara

ALIWAL

Kotli

STEDMAN

GODBY

HICKS

PENNY & WHEELER

McDOWELL

WILSON

Poorain

Entrenchments

Bhundri

▨ BRITISH CAVALRY

☐ BRITISH INFANTRY

◼ SIKHS

Brig. Cureton Com. Cavalry Division

Brig. Stedman { 1st & 5th Native Cavalry
               Body Guard & Shekawati Cavalry

Brig. McDowell { 16th Lancers
                3rd Light Native Cavalry
                4th Irregular Cavalry

Brig. Godby { 47th Native Infantry
             Nusseree Gurkhas

Brig. Hicks { 31st Regiment
             24th Native Infantry
             36th Native Infantry

Brig. Wheeler { 50th Regiment
               48th Native Infantry
               Sirmoor Gurkhas

Brig. Penny

Brig. Wilson { 53rd Regiment
              30th Native Infantry
              Shekawati Brigade

support them by charging through the smoke and jumping the earthworks into the 7 gun battery. Abandoning their guns, the Sikh artillerymen ran to the shelter of squares formed by the Regular Sikh Infantry of Avitabile's Brigade, formed up behind the battery and supported by cavalry. The British horsemen broke through the squares, then turned back into their shattered ranks and the conflict broke up into small mêlées between Sikh infantry and cavalry and the 16th Lancers.

The battle was nearly over, the retreating Sikhs being rapidly followed up by Wheeler's and Wilson's Infantry Brigades. Bhundri, held by native irregulars, was stormed and captured by the 53rd Foot, which cleared out the enemy and captured many guns; the Sikh gunners resolutely stood their ground and fought to the death. Two batteries of Horse Artillery harassed the Sikh foot soldiers running in confusion towards the ford by their camp, where Ranjur Singh had 9 pieces unlimbered to cover the ford; but these guns only fired once before they were overrun by the pursuers. The fugitives tried to escape across the river, losing stores, camp-baggage, supplies and all their 67 guns.

This is one of the most difficult battles to reconstruct realistically so that it culminates in the historical British victory. There are few sets of wargames rules that will allow an army to defeat a force almost twice its size, yet here the enemy is not only stronger in men and guns, but they are entrenched behind earthworks and in houses. The reconstruction must be slanted to balance these inequalities, so that the British can win; but the Sikhs must be given a chance of victory.

Rating of Commanders and Observations

Sir Harry Smith showed himself to be an 'above average' commander, whereas his opponent, Ranjur Singh, displayed 'below average' ability; Cureton, the British cavalry commander, was also 'above average' and the Brigade commanders displayed exceptionally high qualities. It is extremely difficult for the British to 'destroy' the Sikh Army, as in a normal wargame, so the objective could be the Sikhs' need to maintain their lines of retreat over the two fords crossing the River Sutlej; thus the British commander will be required to manoeuvre so that he splits the Sikh force and herds it away from the vital river crossings. When the Sikh commander finds himself unable to prevent this, he will be forced to withdraw and concede the battle.

The Sikh force was surprised by the British advance, for it was marching out of its overnight encampment when Sir Harry Smith arrived on the horizon, and, lacking time to transport his entire army across the river to safety, Ranjur Singh had to send them back to man the earthworks they had thrown up on the previous night. To make the Sikh Army flow back in separate bodies, give their commander an 'order of march' away from his overnight camp, and mark on the map and on a Time Chart the progress of the first unit, then the second and so on, until the British force arrives (on the map). It is possible to ascertain the location of the Sikh units in relation to their defence works. On turning to man their position, the rearmost units will march back to Aliwal, each unit 'peeling-off' to conform to other units until they form a continuous defensive line. Simultaneously, Sir Harry Smith is deploying his force into battle order far enough from the Sikh positions to take a minimum of artillery fire.

A successful reconstruction will depend upon local rules, together with the varying morale and fighting qualities of

(*a*) European (British) infantry and cavalry,
(*b*) Native infantry and cavalry regiments of the British East India Company, officered by Europeans,
(*c*) Sikh Regular infantry and artillery, well trained on Western lines by European mercenary officers, and
(*d*) Irregular Native hillmen and cavalry.

The British-officered Native infantry regiments had been overawed by the martial Sikhs in previous battles, and this must show in their morale and fighting qualities. The Sikhs were brave, well trained and disciplined, and their artillerymen were

VALUES FOR CLASHES BETWEEN DIFFERENT FORMS OF TROOPS

ARM against	European lancers	European sabres	European infantry	Light infantry	Gunners	Native lancers	Native sabres	Native infantry	Tribes-men	Against a square
European lancers	1	2	3	3	4	2	3	4	5	1
European sabres	1	1	2	2	3	2	2	3	4	0
European infantry	0	1	2	2	3	11	1	3	2	1
European lt infantry	0	1	1	1	2	0	1	2	1	1
European gunners	0	0	1	1	1	0	0	2	1	0
Native lancers	1	1	2	2	3	2	3	3	3	1
Native sabres	0	1	1	1	2	2	3	4	2	0
Native infantry	0	0	1	1	1	0	1	2	1	0
Tribesmen	0	1	2	2	3	1	1	3	2	0

prepared to stand by their guns to the death. A proportion of their army at Aliwal consisted of Irregular soldiers—battalions of hillmen and masses of Irregular light cavalry—which were not particularly reliable when facing British troops. In mêlées all types of troops can be given a figure-valuation that varies according to the type of opponents they encounter. To these fighting figures must be added or subtracted other values—the morale of the attacker and defender; whether the attack is in flank, rear or downhill; whether the enemy are outnumbered or behind earthworks, etc. A suggested table is shown on preceding page.

Manipulation of morale status for varying types of troops is a most effective means of reconstructing any wargame between unequal numbers and types of forces. For example, native tribesmen under artillery fire or threat of cavalry attack must check their morale state, with prescribed penalties for these situations (which probably will adversely affect them). The resulting morale level may well cause them to move out of danger or even flee.

Horse Artillery galloping up to within close-range of the enemy, unlimbering, firing and then limbering up and retiring can be simulated on the wargames table by giving Horse Artillery in action the same move-distance as a Light Cavalry charge-move; then penalise them an eighth of their move-distance for the time it takes them to unlimber and another eighth to limber up. With a 24 in move, therefore, a horse gun can limber up in its original firing position (3 in) move forward 9 in, unlimber (3 in), fire, limber up (3 in) and pull out a further 6 in.

In simulating the cavalry actions of this battle it will be necessary to reproduce concealment through smoke and dust (see p 21).

Construction of Terrain
It is necessary to bring the villages of Bhundri and Aliwal and the Sikh entrenchments nearer the river, with the fords positioned on the wargames table so as to form objectives. The River

G

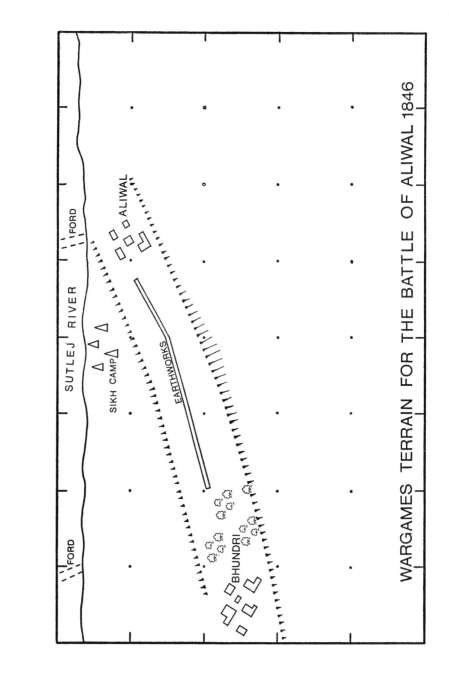

WARGAMES TERRAIN FOR THE BATTLE OF ALIWAL 1846

SUTLEJ RIVER

FORD

FORD

ALIWAL

SIKH CAMP

EARTHWORKS

BHUNDRI

Sutlej can run right along the Sikh baseline to save it taking up valuable wargaming space. The terrain is completely flat, except where a low ridge runs right across the table between the villages. This can be formed by a table-length 'ridge' of books, wood, plastic tiles, etc, under a cloth.

10
The Battle of Wilson's Creek
10 August 1861

THE AMERICAN CIVIL WAR had just started, and General Nathaniel Lyon was fighting to keep the State of Missouri in the Union. The force raised by the Confederates outnumbered his, so Lyon hoped to strike a quick blow before the enemy had concentrated all their troops; realising that he had to retreat, he sought to counter the enemy superiority in cavalry, which made it difficult for him to disengage, by attacking their camp on Wilson's Creek, a meandering stream some 10 miles from Springfield. Plans were made for the Union force to make a night march on 9/10 August and attack the enemy from the north at dawn, while Colonel Sigel's flanking force, making a long detour by a side road, attacked from the south.

The armies of both sides were similar in composition to those that had fought at Bull Run a few weeks earlier, with a few well disciplined and drilled Regular units but mostly ill trained and badly disciplined volunteers. Thus, Lyon's force of 4,282 men and 10 guns was made up of 2 small battalions of Regular infantry, 4 or 5 companies of Regular cavalry, 3 batteries of Regular artillery, 3 semi-Irregular Missouri Regiments (recruited from Germans in the St Louis area), 2 Volunteer Infantry Regiments from Kansas, and 1 Volunteer Infantry Regiment from Iowa. Sigel's flanking force of 1,118 men and 6 guns consisted of 8 companies of the 3rd Missouri Volunteers, 9 companies of the 5th Missouri Volunteers and 2 companies of cavalry (121 troopers).

Under General Ben McCulloch, the Confederate position

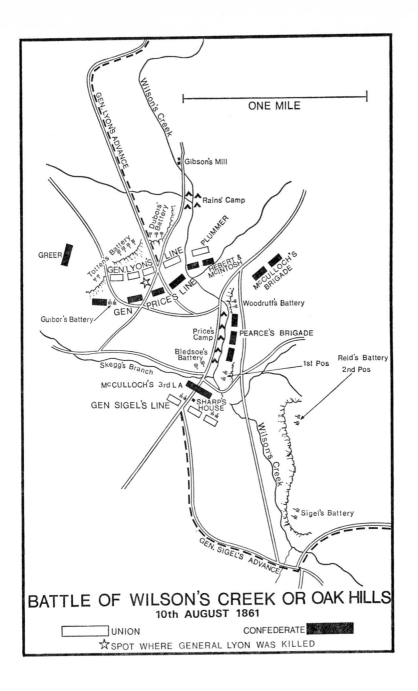

ONE MILE

Gibson's Mill

Rains' Camp

Dubois' Battery

PLUMMER

GREER

Totten's Battery

GEN. LYON'S LINE

HEBERT & McINTOSH

McCULLOCH'S BRIGADE

PRICE'S LINE

GEN

Woodruff's Battery

Guibor's Battery

Price's Camp

PEARCE'S BRIGADE

Bledsoe's Battery

Skegg's Branch

1st Pos

Reid's Battery
2nd Pos

McCULLOCH'S 3rd L A

GEN SIGEL'S LINE

SHARP'S HOUSE

Wilson's Creek

Wilsons Creek

GEN LYONS ADVANCE

GEN. SIGEL'S ADVANCE

Sigel's Battery

BATTLE OF WILSON'S CREEK OR OAK HILLS
10th AUGUST 1861

UNION   CONFEDERATE

☆SPOT WHERE GENERAL LYON WAS KILLED

was such that the sturdy Missouri State Guard, under Major-General Sterling Price, with its 5 brigades (commanded by Slack, Clark, Parsons, MacBride and Rains) was placed to receive Lyon's attack. On the bluffs east of Wilson's Creek were Hebert's 3rd Louisiana Regiment of infantry, well drilled and disciplined, and McIntosh's Arkansas Regiment. Further south were Pearce's 3 regiments of Arkansas Infantry with 2 batteries, while other troops under Greer, Churchill and Major were positioned along the valley. The force totalled 11,500 men, of whom perhaps 6,000 or 7,000 were in semi-fighting trim and participated in the battle. There were 15 guns.

In the area where most of the fighting took place, Wilson's Creek was fairly deep, with rough, steep and rather high banks, making fording difficult. On the left the hills were sheer like bluffs; on the right or western bank the ground was a succession of broken ridges, covered with trees and a stunted growth of scrub oaks with dense foliage. forming an almost impenetrable tangle in places. Rough ravines and deep gullies cut up the surface.

Led by Plummer's battalion of Regular infantry, Lyon's force struck Rains' camp at 4 am, the escaping Confederates warning their main force of the impending attack. The Union force advanced $1\frac{1}{2}$ miles until its skirmishers met the Confederate skirmish line and pushed them back to the ridge where the Confederates had formed their line. On the left Plummer became separated from the main body by a deep ravine and swampy ground; his men entered the cornfield which lay beyond to take on, and be checked by, a large force of the Louisiana Regiment. Then Dubois' battery disordered the Confederates in the cornfield, so that Plummer was able to draw off in good order. On the ridge both sides fought with great bravery and determination in ranks three or four deep at ranges of 30–40 yd, with the muzzle-loaders doing great execution in an action that raged for more than an hour. Each side gave ground in turn, until at last the Confederates ceased coming forward, and a lull ensued.

At 5.30 am the sound of battle was briefly heard in the rear of the Confederates facing Lyon; then it died away, was renewed and finally ceased altogether. It was Sigel who, when he heard the firing from Lyon's action, had opened fire with his 4 guns, placed on the hill overlooking the enemy camp, then pushed his infantry across the Creek and entered the lower camp, the enemy fleeing before him as he advanced against slight resistance to Sharp's house, where the Union force formed line. McCulloch gathered a force and attacked them, to be mistaken by Sigel's men for friends, the confusion being intensified by the enfilading fire of Reid's battery east of the Creek. In the fight that followed Sigel's force was completely routed and by 10 o'clock he was out of the fight.

Conscious that a third of his men were down, that Sigel's attack had failed and, while he had no reinforcements, the rebels in front of him seemed to have an unending supply of men, Lyon was heard to mutter: 'I fear the day is lost'. Then Lyon was killed leading an attack, and Major Samuel Sturgis assumed command. The death of their leader was not known to the Union troops, who fought manfully for the next hour in repelling repeated frontal attacks and attempts to turn their right flank. Both Union Regulars and Volunteers maintained their position against repeated Confederate charges. Despite the overwhelming numbers of Confederates on the field, the contest was evenly balanced until Dubois' Union battery moved up to pour a murderous fire into the enemy's right flank, making them recoil along their whole front. The Confederates drew off and the battle seemed to have ended.

At 11.30 am, informed of Sigel's rout, and conscious of his depleted and exhausted forces and a shortage of ammunition, Sturgis decided to withdraw to Springfield. Covered by artillery, he left the field, watched by the relieved Confederates, whose ammunition was exhausted, as were their men, who, lacking discipline, were unsuitable for the task of pursuing the Federals. As Dupuy says: 'Although this was a tactical victory for the South, Lyon's aggressive operations had saved Missouri for the Union'.

The losses for both sides were heavy: *Union*, 223 killed, 721 wounded, 291 missing, totalling 1,235; *Confederate*, 265 killed, 800 wounded, 30 missing, totalling 1,095.

Ideally, this reconstruction should be fought as two actions to allow for the possible effect of Sigel's attack on the main action, when the wargame map must be divided into the two areas of battle and each scaled up to fill a full-size wargames table. When Sigel is pressing forward into the Confederate position, or he has been turned back and his opponents return to fight off Lyon, the battle is transferred to a single table. If it is fought as one battle, Sigel's attack must be very carefully timed on the Time Chart; it will be moved on the map, to come on to the table $1\frac{1}{2}$ hours after the main action has begun.

Rating of Commanders and Observations
Lyon is 'above average', McCulloch and Price 'average' and Sigel 'below average'.

Lyon's high rating helps to balance the Confederate numerical superiority, as will the varying fighting qualities and morale of Regulars and Militia, with the latter showing variable and widely fluctuating standards in both aspects. Although the Confederates seemingly had about 11,500 men, apparently only some 6–7,000 took part in the battle. Therefore the remainder must be occupied elsewhere, posted in outlying villages and areas, and a Military Possibility can bring them either marching to the scene of the action (on the map at scaled move-rate) or cause them to remain where they are for fear of being attacked. The former Possibility would provide an alternative objective, in that Lyon has to defeat the Confederates facing him before the arrival of reinforcements.

The morale of the Regulars should enable them to remain steady while their Militia comrades, although being given the chance to display the same courage, are liable to give way under stress. The factors governing firing and hand-to-hand fighting must be loaded in the Regulars' favour, to simulate the steadiness under fire that results from discipline and training.

At the onset of the battle the Confederates fleeing from Rains' camp warned the main force of the impending attack, though the sound of firing would have had the same result—a Military Possibility could consist of a strong wind blowing away from the camp and carrying away the sound of firing (see chapter called 'Weather in Wargames' in *Advanced Wargames*). The Time Chart must show the time of the attack on Rains' camp, delay for possible resistance, the time (measured in distance) taken by the escaping soldiers to reach the main camp, and the delay while the alarmed soldiers are roused and commanders issue their orders. Meanwhile, Lyon's force is steadily advancing.

Surprise and confusion will undoubtedly cause the Confederates to form up hastily, since few were well trained or disciplined soldiers. This initial advantage for Lyon's smaller force can be simulated by allotting a specific period of time for the Confederates to form up and move forward to the battle line; the shorter this period, the less prepared is the battle line, with a subsequent depreciation in its fighting ability and morale.

The scaled-down armies of wargames' tables reduce the ranks from 3 or 4 to 2 deep, but, even so, double-rank firing will be extremely effective. A marked feature of the American Civil War was regiments running from the firing line, to be rallied 100 yd back and to return and fight bravely. Hence, the battle line in this encounter will show units dropping out, moving up to fill gaps, and being outflanked; and it will test the stoutheartedness of their commander in most realistic fashion.

A Military Possibility might arise from the sound of Sigel's attack in their rear affecting the morale of the troops holding off Lyon's force. This could be represented by Chance Cards or a personal morale-check on individual commanders such as McCulloch and Price, or even unit commanders.

In the early morning mist and smoke, Sigel's men mistook oncoming Confederates for friends. This situation is probably best handled by the wargamer representing Sigel drawing a Chance Card, which might bear an instruction such as 'Hold fire until oncoming troops within 6 in as they may be friends' or other injunctions, ranging from the best from his point of view to the worst. Plummer's regiment on the Federal left could take

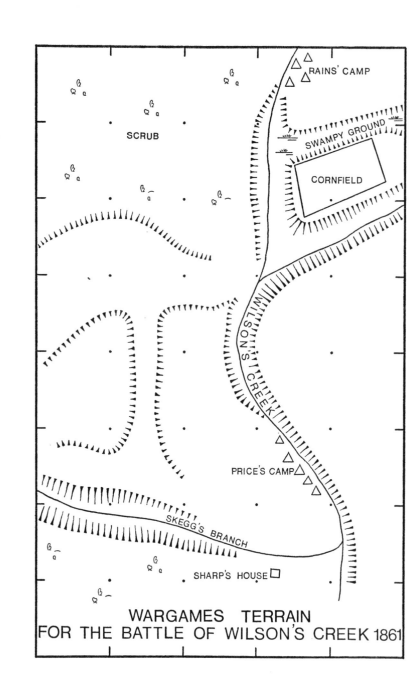

RAINS' CAMP

SWAMPY GROUND

CORNFIELD

SCRUB

WILSON'S CREEK

PRICE'S CAMP

SKEGG'S BRANCH

SHARP'S HOUSE

WARGAMES TERRAIN
FOR THE BATTLE OF WILSON'S CREEK 1861

advantage of a Military Possibility to act differently, although it seems that, together with Dubois' battery, his force held the attention of a much larger group of Confederates than itself.

Lyon's death was kept from his army, but a Military Possibility might make nearby units aware of it. The small Federal army was primarily held together by the strong personality and enthusiasm of Lyon, so that his death could have an adverse effect upon its morale—this must be reflected if necessary. Historically, Lyon's successor Sturgis, almost out of ammunition, drew off what was left of his exhausted force and withdrew to Springfield. Thus, the ability of the Federal force to withdraw from the field in good order (providing they have administered proportionately equal casualties to the Confederates) should provide them with some satisfaction, though the tactical victory will go to the South. Perhaps more than any other of our battles, this one is likely to satisfy both commanders.

Construction of Terrain
It is necessary to compress the salient features of the field, and modify its undulations, vegetation and other topographical features so that it forms a practical battlefield for model soldiers to fight upon. The southern aspects (where Sigel attacked) must be brought nearer the area of the main battle line. Almost the entire right-hand third of the table forms a plateau steeply sloping down to the creek on the left and to the creek branch running off parallel to the cornfield; there must be a ravine with swampy ground immediately below Rain's camp. On the left of the table some irregular scrub-covered ridges will form the main area of fighting between Lyon and the Confederates. Further south, Skegg's Branch was bordered by fairly steep banks, with scrub-covered ground in the vicinity.

11

The Battle of the Little Big Horn

25 June 1876

ONE OF the last battles in which the American Indians resisted the white man, this consisted of two entirely separate actions. First, Major Reno's troops in the valley, retreating before an overwhelming force of Indians, were joined by Captain Benteen's force, and defended themselves on the bluffs across the river until late on the following day. In the other action, fought nearly 5 miles away, 5 troops of the 7th Cavalry under George Custer were overwhelmed and wiped out in less than an hour by a large Indian force.

Lt-Colonel George A. Custer took his 7th Cavalry, consisting of about 600 soldiers, 44 Indian scouts and 20 guides, in pursuit of a large Indian force, with the idea of compelling them to fight to avoid being trapped between converging columns. On sighting the Indians, Custer impetuously decided to attack them, without realising that the Indian force consisted of about 5,000 warriors, the majority Sioux and Cheyenne under Crazy Horse, Sitting Bull and Gall.

At about noon on 25 June Custer divided his command into three battalions—Company's A, G and M under Major Marcus A. Reno; H, D and K under Captain Frederick W. Benteen; and C, E, F, I and L under his own immediate command; with B Company protecting the pack train, which was to follow the main column as closely as possible. Benteen scouted left of the trail while Custer and Reno proceeded along opposite banks of a small creek towards the Little Big Horn valley. Reno, with 112 men and 20 scouts, was told to cross the river and attack

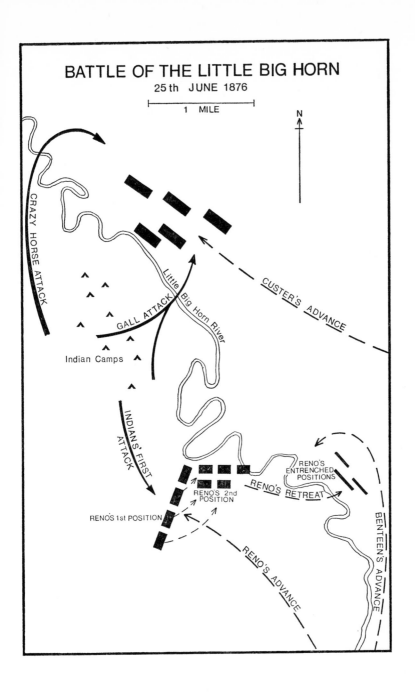

BATTLE OF THE LITTLE BIG HORN

25th JUNE 1876

1 MILE

N

CRAZY HORSE ATTACK

CUSTER'S ADVANCE

Little Big Horn River

GALL ATTACK

Indian Camps

INDIANS' FIRST ATTACK

RENO'S ENTRENCHED POSITIONS

RENO'S RETREAT

RENO'S 2nd POSITION

RENO'S 1st POSITION

BENTEEN'S ADVANCE

RENO'S ADVANCE

a camp on the west bank, while Custer turned right to support him in the river bottom by suddenly appearing at the lower end of the Indian camp and attacking its flank and rear.

So, at 2.30 pm Reno crossed to the west of the Little Big Horn River and advanced down the valley towards the Indian camp, but soon he was attacked by hordes of Indian warriors. His men dismounted and deployed in a skirmishing line, fighting on foot for half an hour, until the increasing numbers of Indians forced them to make a defensive stand in a timber thicket. Reno ordered his men to mount and retreat to the bluffs, a move which became a panicky flight as the Indians attacked the soldiers crossing the river, where they lost 3 officers and 40 men killed, wounded or seeking cover in the brush. Across the river, on the east side, the survivors took up a defensive position on the top of a hill, but the Indians did not follow.

Leaving Reno, the mass of Indians started after Custer's column, which had advanced to the junction of two ravines just below a spring, where Custer dismounted two companies, under Keogh and Calhoun, to fight on a knoll. The remaining three mounted companies continued along the ridge and then dismounted to occupy for about ¾ mile a line along the first considerable ridge beyond the river.

Gall and his warriors were the first to engage Custer, before another large party under Crazy Horse moved from the lower part of the encampment up a ridge to cut Custer off from the village. Outnumbering Custer's men by as many as 20 to 1, the Indians did not rush madly to the attack, but fought on foot and from prone positions, as did the Americans. Large numbers of Indians wriggled along gullies and hid behind knolls all round the troops—it was a terrain that lent itself to this style of fighting. Many of the troopers' horses were shot to make breastworks, but only 225 cavalrymen, their ammunition running out, could not hold out against as many as 5,000 warriors. The action is believed to have lasted about an hour, and it is said to have ended when the Indians stampeded the soldiers' horses, so that they lost the extra ammunition in their saddlebags. This was immediately followed by a concerted Indian attack, so successfully and swiftly carried out that within minutes not an American trooper remained alive. There was no final charge on horseback as portrayed in writings and paintings.

Just after 3 o'clock Benteen and his 3 companies pushed forward and joined forces with Reno's depleted and shaken command in defensive positions on the high bluffs. An hour later Captain McDougall came up with his company and the train of pack mules.

Believing that Custer might need assistance, Captain Weir took Company D, followed by Benteen and the other two companies of his battalion, to a point about $1\frac{1}{2}$ miles to the north-east, when large numbers of Indians began a frantic ride to cut them off. In a few minutes the force was so threatened that the troopers dismounted and prepared to fight on foot. Realising that the position from which they had set out was the stronger, and to make the force a more compact body, Weir ordered a withdrawal, and no sooner had the force reached its original position on the bluffs than Indians appeared from everywhere and heavy firing was exchanged until dark, when the warriors withdrew to the valley. During the night the troopers dug trenches and made barricades of food boxes.

The Indians resumed the attack at dawn and fighting continued throughout the morning and afternoon, when the warriors withdrew, leaving a small group to harass the cavalrymen. Late in the afternoon the Indians fired the grass in the valley to cover the departure of the entire Indian encampment. During the 2 days of fighting, Reno and Benteen lost 32 men killed and 44 wounded. Indian losses in the entire action have never been precisely known, but published figures vary from 30 to 300.

As Custer's and Reno's battles were well separated, this is another reconstruction that lends itself to being fought as two actions, each bearing upon the other in the long run. The only other choice is to telescope them together and, with wargamer's licence, fight them on the same table at the same time, modifying the reconstruction by

(a) telescoping the two fields of action,
(b) having Custer try to rejoin Reno, on realising the odds against him, and
(c) allowing the Indians to split to fight Custer and Reno, with the latter's defensive position drawing off enough Indians to give Custer a chance.

In fact, the attack on Reno ceased when the Indians started after Custer's column, which more or less rules out the only, though remote, chance of reversing the decision. Historically, when Custer's men were slain, the Indians returned to attack Reno; in a table-top simulation, only if and when either Custer or Reno are wiped out, will the Indians concentrate on the remaining force, so that both actions occur simultaneously.

Custer's initial objective was to destroy the Indians and their encampment, but his advance turned into a fight for survival, giving rise to an alternative objective—that if Custer and Reno put up a prolonged resistance, the Indians should be allowed a given percentage of losses or a time limit and, when either of these is reached, the Indians will have to withdraw.

Ratings of Commanders and Observations

Custer rates 'below average' and the Indian chiefs 'average'.

This reconstruction presents a peculiar problem: if the heavily outnumbered 7th Cavalry is to have *any* chance of success (if not the reconstruction becomes purely historical), strong factors must give the Americans parity with the Indians. This is difficult, because the Indians were as good as if not better fighters in this environment than the Americans; and their morale was certainly as good, probably higher. Thus, the usual 'differing morale' method of handling numerical inequalities cannot prevail here, and disparity in numbers must be tackled through the firepower of the 7th Cavalry when opposed to the relatively poor trade muskets and even bows and arrows of the Indians. The ·45–70 Springfield carbine, carried by troopers in addition to their Colt revolvers, fired a hard hitting bullet with a flat trajectory, and had a much greater range than any weapons carried by the Indians. Reflect this by allowing the cavalrymen greater range and the ability to fire much more frequently (perhaps three times per game-move), though rapid fire causes them to run out of ammunition quickly.

An alternative method of handling the numerical disparity, though this method precludes an accurate reconstruction, is to precede the actual game by a period of map-moving, giving Custer the opportunity of manoeuvring advantageously so as to use his superior firepower against Indian groups in detail rather than *en masse*.

A Military Possibility can be introduced to prevent Custer's horses being stampeded along with the reserve ammunition. If so, the possibility of the Americans running out of ammunition during the course of the battle, due to the rapid expenditure of the cavalry carbine, should be reflected in subsequent events.

Reno's party is said to have scrambled back across the river in confusion, which, in a wargame, will have to be put down to low morale; and on reaching the bluffs they will have to be rallied. If they do not rally, they will flee the field, taking with them all the realism of the reconstruction. If they are not considered to have been routed, their hasty withdrawal has to be tacitly accepted as part of the wargames action, and some 'local' rules devised to cover it.

A Time Chart can be a great help when reconstructing this action, tying in the various movements of Custer, Reno, Benteen, Weir and the Indians.

Construction of Terrain

The battle is best fought along the length of the table, with the scene of Reno's action at the bottom, where the lower eighth of the table is raised to form a plateau with steep slopes descending to the river. A series of scrub-covered ridges and valleys will

H

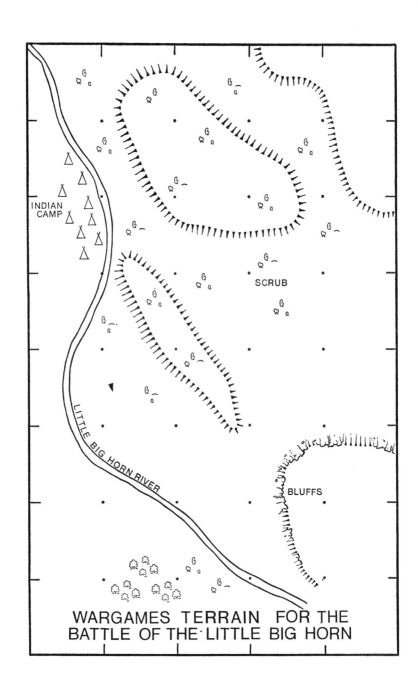

WARGAMES TERRAIN FOR THE
BATTLE OF THE LITTLE BIG HORN

form the top half of the table, representing the ground on which Custer fought. It is not a difficult terrain to make, and the ridges and hills can either be formed of setpiece features or by moulding a cloth over shapes beneath it.

12
The Battle of Modder River
28 November 1899

THIS BATTLE was fought during the Second Boer War between the British 1st Division, commanded by General Lord Methuen, and Transvaal and Orange Free State Boer commandos under Generals Cronje and De La Rey.

Methuen's column, moving to the relief of Kimberley, had to cross the Modder River, which, like many South African rivers, ran through a miniature canyon about 30 ft below the level of the veld, where a long line of bushes and trees marked its east to west course. Methuen had no map of the area nor any details of the surrounding country, and discounted information about a weakly held drift lower down the river. He did not expect much opposition, as he was convinced that only about 400 Boers had been left behind to delay his force.

Anticipating the usual British frontal attack, the Boers had constructed elaborate defences: entrenchments masked by shrubs and brushwood stretched for 5 miles along both banks giving Boers on the southern bank a clear field of fire across the coverless veld, which sloped gently towards them. Their smokeless powder offered no visible target, so that the British artillery would almost certainly range on buildings on the north bank, where 7 field guns were positioned. There was also a heavy gun (probably a 100-pdr) on high ground to the rear. Numerous artillery emplacements were constructed to delude the British into thinking they had silenced the guns when they had, in fact, been moved from one place to another. Whitewashed stones placed on the veld gave exact ranges both to the Boer gunners and riflemen. On the Tewee Rivier, a tongue of land between the Riet and the Modder, they dug-in a 'pom-pom' and two field guns, and they also positioned several Maxims and machine guns along their front.

In short lengths, the trenches were irregularly aligned, with their parapets concealed by rocks and bushes. Each held about 6 men. The farmhouses and buildings were converted into strongpoints. The Transvaalers were on the left of the position while the Orange Free Staters were on the right; it is thought that De La Rey positioned the latter with their backs to the river so that they would find it harder to break off the fight than they had done in their previous two battles.

About 6.30 am the British 18th and 75th Field Batteries unlimbered on the right and opened fire at a range of 4,500 yd. Boer guns replied with a flash and a faint film of blue-white smoke instantly dissolving in the air. This long-range firing continued for some time, until the Boer guns ceased firing to give the impression that they had been silenced, and that their small rearguard was falling back. At that time the 9th Lancers, patrolling about a mile from the river, withdrew under a sharp fire to the extreme right of the line, where they took no further part in the battle.

At the regulation 5 paces interval, the Guards moved majestically down the slope, Scots Guards on the right, 3rd Grenadiers and 2nd Coldstream echeloned to the left, and 1st Coldstream in the rear. Descending the smooth grassy slope leading gently down to the river, they got to within 800 yd of the enemy's trenches when suddenly from along the whole extent of the Boer front came musketry interspersed with Maxim fire. The advancing British fell flat to the ground, to be pinned down under a pitiless sun for the remainder of the day, their slightest movement attracting a storm of bullets.

The 1st Coldstream Guards, extending to the right to cover and support the Scots Guards, found to their surprise that they were halted by the Riet River running south to north, a fact completely unknown to the mapless Methuen. Nor was he aware that a few hundred yards back was Bosman's Drift, where a

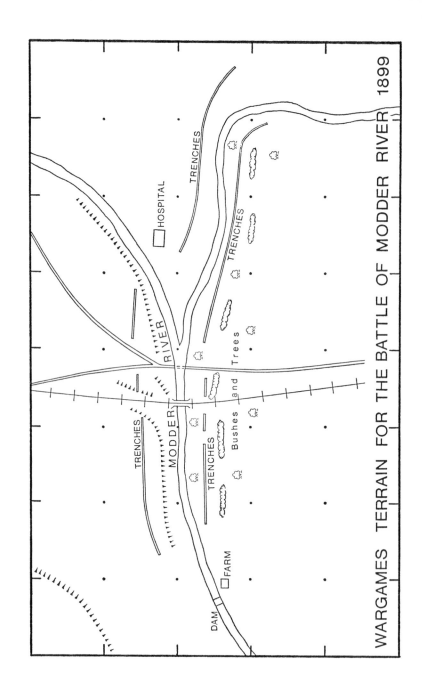

WARGAMES TERRAIN FOR THE BATTLE OF MODDER RIVER 1899

crossing could have been made in strength, taking the entire Boer position on the Modder in reverse. A small party of the Coldstream Guards managed to cross the river, but at a place that was obviously impossible for large numbers of men. The battalion dug themselves in so that the entire Brigade of Guards were out of the battle, and it was not yet 8 am. With the halting of the attack, the artillery, in positions less than 1,300 yd from the Boer trenches, spent the day covering the north bank with shrapnel.

Far away to the left the 9th Brigade pressed forward, Northumberland Fusiliers on the right, King's Own Yorkshire Light Infantry in the centre, North Lancashires on the left, and the Argyll and Sutherland Highlanders in support of the right and centre. The KOYLI and Lancashires stormed a farmhouse in a kraal just to the south of the dam, getting within charging distance of the buildings by moving along an unseen and unprotected shallow ditch running down to a small clump of trees by the river bank. The Boers in the farm and in the trenches on either side were unable to fire on them for fear of hitting each other. Recklessly leading an attempt to cross the river, Lord Methuen himself was wounded and compelled to hand over his command to Brigadier Colville. Led by General Pole-Carew, the KOYLI, one by one, clambered across the river on the dam under a heavy fire, until 400 men had formed up on the other side; then they pushed along the north bank hoping to take the enemy in flank. Unfortunately, they were mistaken for Boers and fired upon by their own artillery, which compelled them to fall back. However, their appearance on the north bank so alarmed the lukewarm Orange Free Staters that at 2 pm, a large number of them mounted and rode off; at 4 pm there was something resembling a general stampede as the Boers retired along the deep river bed out of sight of the British, who did not realise what was occurring.

Earlier in the afternoon the 62nd Battery, after dashing 62 miles in 28 hours, arrived and immediately went into action, together with the rest of the artillery, as the centre of the battle moved towards the left of the position. An attempt to resume the Guards Brigade attack failed in the face of heavy fire, so it was decided that they should move out under cover of darkness, cross the river to the left and storm the enemy position.

In the early evening, after 8 hours of continual fighting, the remainder of the enemy retreated, unknown to the British, leaving behind their guns and many of their wounded. Later that night they mustered up courage to return and collect them. The British bivouacked on the field where they had fought and next morning made an unopposed crossing of the river to take the abandoned enemy position.

British casualties were 466 men killed, 20 officers and 393 men wounded, and the Boers lost 60 killed and 300 wounded, in what was a military farewell to the nineteenth century, in that the bewildered British soldier found himself severely punished through being tactically untrained to tackle a dug-in enemy possessing a high rate of firepower.

As the Boers have numerical parity with the British, plus superb cover, how can the British enter this battle with even the slightest chance of success? Here is a selection of Military Possibilities that might affect the course and eventual result of the action:

1. The OFS commandos are given a lower morale standard than their Transvaal comrades so that, under stress, they break and retire.

2. The British force can discover the existence of Bosman's Drift lower down the Modder River and cross there, turning the Boer flank. However, while this move may provide an interesting wargame, it takes all reality from the battle.

3. Pole-Carew's success on the left flank can be exploited.

4. The Coldstreams' abortive crossing on the right could be made a successful venture, again turning the Boer position.

5. The Boers' (particularly the Free Staters') lack of experience and dislike of being on the receiving end of artillery can be reflected by morale penalties when under fire, so that they might break at any point, particularly when the heavy guns of the 62nd Battery arrive.

6. The Boers' defensive dispositions were all made in anticipation of a British frontal attack, and De La Rey had early nervous moments when the Guards moved towards the flank before veering centrewards. A flank attack would have caused De La Rey to alter his dispositions completely, under artillery fire, or else withdraw. Again, however, this detracts from the realistic reconstruction of the battle.

Rating of Commanders and Observations
De La Rey stood out in comparison with the other leaders but to make him an 'above average' commander and Cronje 'average' is to overweigh the Boers' already strong hand. Therefore, both De La Rey and Cronje should be rated as 'average', and Methuen 'below average', but Pole-Carew, on the British left, could be rated as 'average'. Or it might be more realistic to rate De La Rey and Pole-Carew as 'above average', and Cronje and Methuen 'average'.

In spite of their inability to move forward, the British Regular soldiers cannot be considered to be low in morale or in fighting ability, if only because they possessed training and discipline superior to that of the Boers. The penalty for a low morale rating for the British soldier could be for them to 'go to ground' when they come under heavy Boer fire, when they have three choices:

1. Once down they take no casualties but cannot fire back.
2. Remaining on the ground, they can return fire but take half casualties.
3. They can fire on their feet and move their full distance, or they can move and fire with half effect, or they can charge-move and make contact with the entrenched Boers.

The condition governing each choice rests upon their state of morale under fire, decided by throwing a dice for each group and deducting 1 from it (a) for *any* losses, (b) if they come under fire from a group of men approximately twice their strength, (c) if they are under artillery fire, or (d) if they are under fire from flank or rear.

To be able to rise to their feet and move or fire, a group needs to total 3; to stay down and return fire the group needs to total 2; to stay down without casualties and without returning fire is their normal basic reaction, and is what they will do if unable to make the required totals.

The Transvaalers' morale will benefit from being behind cover but they should be rated as equal in morale and fighting ability to the British Regulars in fire-fights but slightly lower in mêlées because of their fear of the bayonet. Having shown a disinclination to fight at the earlier battles of Graspan and Belmont, the Orange Free Staters are regarded with some contempt by the Transvaalers, and it might be a reasonable Military Possibility to insert a built-in distrust in the game rules so that their actions are not well coordinated. The lower quality of the Free State Troops should be reflected in their morale and fighting ability.

Having received a grim lesson from British Lancers at Elandslaagte, the Boers greatly feared the lance, yet the 9th Lancers played no part in the Battle of Modder River. A Military Possibility could permit them to move from the right flank if desired.

The British artillery's firing on its own men on the left can be controlled by Chance Cards or, as the firing will be 'off-table' map-shooting, a certain allowance can be accepted for inaccuracy of aim. If it should happen that the Biritsh soldiers are hit, their morale must immediately be checked, with a 'distraction' factor to allow for their being fired on by their own side.

One realistic way of refighting this battle is to give the wargamer representing Methuen only the details that were known to the British commander in 1899—400 Boers forming a rearguard—only to discover, as did Methuen, that the situation is very different. As Methuen had no map of the area and discounted local information, the wargamer playing Methuen must be told that some or all of the information may be false; he has no alternative, therefore, but to make his plans and dispositions on a sketch-map conforming to his own vague idea of the terrain and what he is told that he can see. (This must be done

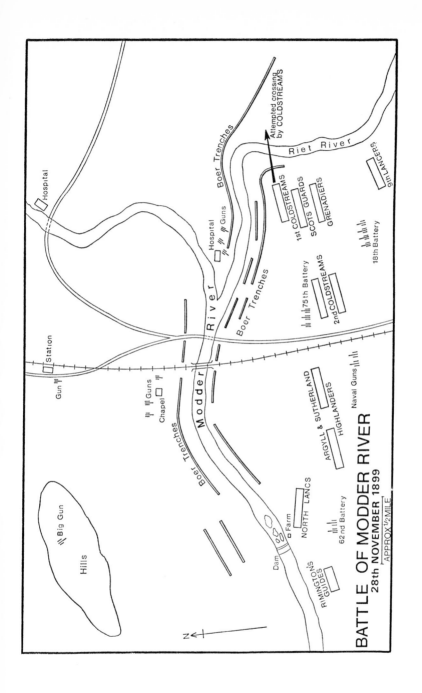

BATTLE OF MODDER RIVER
28th NOVEMBER 1899
APPROX ½ MILE

before he sees the table-top terrain.) The Boers will mark their dispositions on their own accurate map, and the troops from each side will be shown when action demands. Any casualties caused by artillery fire will be deducted on paper, the map being marked accordingly by the Boer commander, and morale checked whenever necessary. He will move on paper as if on the actual table, but any men breaking cover (such as in rout or in changing their position) will have to be revealed on the table.

Construction of Terrain

The Modder River will bisect its width just above centre, thus allowing ample space for the Boer positions in front of the river and for the British main area of action. There must be adequate space on the British left flank for the attack on the farm and the subsequent crossing of the river by the dam, and also sufficient on the right flank in case Military Possibility allows the Coldstream Guards to cross at the shallow point. The ground above the river should be slightly raised, so that the Boers have a field of fire over the heads of their men in the trenches forward of the river. The heavy gun can either fire from the hill at the top left-hand corner of the table or 'off the table' as a 'map-shoot'.

The battle might, in fact, be fought in two parts, as the British front was 3 miles long, and the soldiers at one end had no idea what was happening at the other.

13
The ANZAC Landing at Gallipoli
25 April 1915

AT DAWN 48 small boats crept towards the shore at Ari Burnu; a flare shot up, a warning was shouted and sporadic fire broke out as about 4,000 men of the 1st Australian Division landed in 3 successive waves on a 2,000 yd front. As the men grouped on the beach in the steadily improving light of day, it became obvious that the landing had taken place on the wrong beach – the soldiers had been told they would find a low sandbank, whereas this beach was backed by almost sheer shrub-covered cliffs. Nevertheless, since it had been drilled into the men that everything depended on moving quickly inland, the Australian soldiers scrambled up the steep slope to a small flat area— subsequently known as Plugge's Plateau—just in time to see Turkish soldiers hastily disappearing down precipitous slopes into a vast scrub-covered ravine, later to be named Shrapnel Valley.

In spite of the general confusion caused by the difficult ground and intermixing of units, by 6 am about 4,000 Anzacs were ashore and moving rapidly inland, driving the 700 or so Turks in the area back in disorder towards Baby 700 and Mortar Ridge. By 7 am small detached parties of Australians and New Zealanders had crossed Legge Valley, and the capture of all objectives seemed imminent. But, as the sun climbed high in the clear sky, the battle increased in intensity, with Turkish resistance mounting.

New arrivals splashing ashore at Anzac Cove were sent to Shrapnel Valley to reform and be despatched to various parts of the confused battlefield. Maps were useless and orders based on them could not be carried out: some New Zealanders took over an hour to move from Plugge's Plateau to Russell's Top because the map did not show an impassable 'razor edge' between them; 6 guns of an Indian Mountain Battery were, with incredible effort, dragged on to an exposed position on Plateau 400, where Turkish artillery soon wiped them out. The brief firing of this battery was the only artillery support the infantry received throughout the day, as the Navy lying offshore could not fire for fear of hitting their own troops. By 2 pm more than 12,000 Anzacs were ashore, opposed by about 4,000 Turks, but numbers were a dubious advantage, as the nature of the ground, the confusion caused by landing in the wrong place, and the intermixing of units blunted their striking power.

The ground was very difficult to fight on—covered with arbutus, dwarf holly, oak and stunted pine 3–12 ft high and so thick in places as to be impassable. The scrub-covered rocky sides of Monash Valley were steep and precipitous; at this time of the year it was an arid valley with no exit save the almost perpendicular waterfalls that scored its sides. The most important, and bitter, fighting raged around the narrow ridge—100 yd wide and with a sheer drop on both sides—which connected Plateau 400 with Baby 700 and The Nek. A party of Australians had occupied The Nek by 8 am, a group under Captain Lalor remaining there while Captain Tulloch took the rest of the men over to Baby 700.

Mustapha Kemal, the best of the Turkish commanders, was in charge in this area and, bringing his troops up quickly, he forced Tulloch's small party to the back slopes of Baby 700, the Turks establishing themselves on its Western slopes and infiltrating its seaward flanks, so that it was not long before the Anzacs were forced to regroup just above The Nek.

At 11 am 2 companies arrived to reinforce them and, joining with Lalor's men, they charged the summit of Baby 700 and occupied the position. This summit changed hands no less than 5 times, as the local numerical superiority of the Turks drove the Australians from it and then, as reinforcements rushed up from

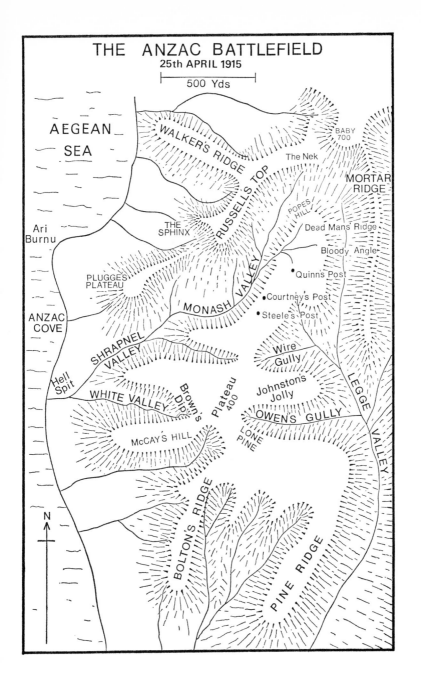

THE ANZAC BATTLEFIELD
25th APRIL 1915

500 Yds

AEGEAN
SEA

WALKER'S RIDGE

BABY
700

The Nek

MORTAR
RIDGE

RUSSELL'S TOP

POPES
HILL

Ari
Burnu

THE
SPHINX

Dead Mans Ridge

Bloody Angle

PLUGGES
PLATEAU

Quinn's Post

MONASH VALLEY

Courtney's Post

Steele's Post

ANZAC
COVE

SHRAPNEL
VALLEY

Wire
Gully

Hell
Spit

Brown's
Dip

Plateau
400

Johnstons
Jolly

WHITE VALLEY

OWEN'S GULLY

LEGGE VALLEY

McCAY'S HILL

LONE
PINE

N

BOLTON'S RIDGE

PINE RIDGE

Anzac Cove, they in their turn drove the Turks back. But by 3 pm the Australian and New Zealand troops, reduced to the fragments of 7 battalions all intermixed and without their familiar officers, were fighting with increasing desperation for the vital high ground above Monash Valley.

Between 4.30 and 5 pm a Turkish counterattack, developing across Legge Valley to Plateau 400 and down the slopes of Baby 700, drove the remaining Anzacs back into positions on either side of Monash Valley and The Nek. Captain Lalor was dead and reinforcements ceased to arrive. The Turks came on in waves, suffering very heavy losses but fighting bravely; neither side took any prisoners and groups of men were cut off and killed to a man in the frenzied and savage conflict.

At dusk the Anzacs were clinging to a series of detached positions at the end of Russell's Top, along the eastern side of Monash Valley and just inland of the crest of Plateau 400. A party of New Zealanders hung on to a vital position just south of the abandoned Nek. Colonel Pope occupied the small rocky eminence at the head of Monash Valley, subsequently known as Pope's Hill and, to his right along the eastern crest of the valley, detachments of men were clinging to exposed positions, subsequently known as Quinn's, Courtney's and Steele's Posts; to the south the Anzacs had been driven almost to the seaward edge of Plateau 400. By 5 pm the Anzacs, fighting for their lives, were being driven back to the sea, with everyone praying for night to fall. In the event, they held on.

Essentially, this will be a wargame fought between small groups of men completely separated from each other by steep hills and precipitous scrub-covered valleys. The success or otherwise of one group will have little immediate effect upon others fighting in the next valley or on the next plateau. Like Pork Chop Hill, this action is a jumbled affair fought desperately between small detached parties of men. It is best to set up a terrain and, with Anzacs and Turks in their right numerical proportion, let the battle fight itself. The objectives are for the Anzacs to hold on and for the Turks to throw them back into the sea. The Anzacs, to win the battle, must hold all or most of those positions they were holding at nightfall on 25 April 1915—at the end of Russell's Top; on the eastern side of Monash Valley; just inland of the crest of Plateau 400; just south of The Nek; Pope's Hill, the small rocky eminence at the head of Monash Valley; the exposed positions subsequently known as Quinn's, Courtney's and Steele's Posts; and the seaward edge of Plateau 400. Each could be allocated a points value, and an agreed total has to be reached for the Anzacs to win.

Rating of Commanders and Observations

As any general commander can exercise little or no influence on this muddled action, the Anzacs can be led by Captains Lalor and Tulloch, although they only represent two out of dozens of officers present. The local effect of Mustapha Kemal must be reflected by an 'above average' classification.

Morale must be calculated for each individual group. Throughout the morale of the Australian and New Zealand troops was first-class. At the beginning the Turkish troops' morale was poor, but, under the influence of Mustapha Kemal and buoyed up by local successes, it should rise on a sort of sliding scale. For the first third of the game Turkish morale could be low, for the second third normal, and for the final third as high as that of the Anzacs; or it can begin at 'average', rising and falling in accordance with local successes.

The battle can either begin with the Anzac groups in position, as far forward as Baby 700, and then fighting grimly to hold their positions while being pushed back by increasing groups of Turks, who arrive at specified times; or with the Anzacs moving from the beaches across the terrain towards the features they historically reached, the movements of both sides being co-ordinated on Time Charts. One method of simulating the inaccurate maps and general confusion is to have the Anzac force controlled by an outside commander, who will be given a deliberately faulty map and prevented from seeing the table-top terrain; he will receive vague messages from the troops fighting forward of him and will order the reinforcing troops forward

I

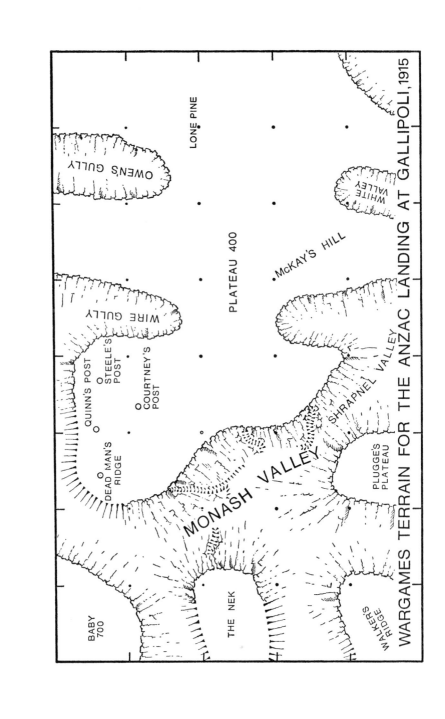

LONE PINE

OWEN'S GULLY

WHITE VALLEY

PLATEAU 400

McKAY'S HILL

WIRE GULLY

QUINN'S POST

STEELE'S POST

COURTNEY'S POST

DEAD MAN'S RIDGE

SHRAPNEL VALLEY

MONASH VALLEY

PLUGGE'S PLATEAU

BABY 700

THE NEK

WALKER'S RIDGE

WARGAMES TERRAIN FOR THE ANZAC LANDING AT GALLIPOLI, 1915

from Anzac Cove by sending written instructions to their commanders but, as his map is inaccurate, those commanders may well be bewildered by his instructions, and the troops will probably fail to arrive at the places at which they are most desperately needed.

Construction of Terrain
This is hard yet easy—only a sand table or moulded plasticine can adequately reproduce the ruggedness of the area yet, because it was so irregular, almost any formation is suitable, providing the named salient features are represented. Moulding a cloth over 'hills' will suffice, providing that the slopes, except where known to be precipitous, are climbable by model soldiers. The whole terrain can be littered with small pebbles, gravel, etc. (In fact, dog-biscuit fragments, when scattered around, give a terrain a remarkable resemblance to rocky ground.) Use lichen moss to represent the profuse thickets, but remember that the model troops have to be able to stand up; the natural features should serve as obstacles to channel troops away from the control of their commander.

One of the few certain occurrences during the action was the destruction of the mountain gun battery on Plateau 400. For the sake of accuracy, this battery should be placed on Plateau 400 and remain there for a designated number of moves, when the battery commander can, by dice throw, Chance Cards, etc, decide whether or not circumstances dictate a move before he is wiped out. The Turkish artillery that destroyed him will be firing 'off-table' on a 'map-shoot', but the mountain battery is permitted to fire at whatever enemy it can see on the table.

14
The Raid on St Nazaire
27/8 March 1942

DURING WORLD WAR II a British commando and Naval raiding party attempted to block the German-held port of St Nazaire, which had the only dock outside Germany large enough to accommodate the two great German battleships *Bismarck* and *Tirpitz*; the intention was to prevent the Germans forming an Atlantic raiding force based on St Nazaire and Brest. Because of its importance, St Nazaire was well defended with coastal and dual-purpose anti-aircraft guns, plus a garrison of 3–4,000. The 6 miles of river approach were negotiated by the attacking vessels moving, at high tide, over the sandbanks in the middle of the estuary. The air raid laid on for 40 minutes before the actual assault, to divert attention, was not pressed home because of low cloud, but it alerted the defenders.

The attacking force consisted of 280 commandos and 350 naval personnel carried in the destroyer *Campbeltown*, 16 motor launches, 1 motor torpedo boat and a motor gunboat.

In the early hours of the morning of 28 March the attacking force made its way unobserved up the estuary to within less than 2 miles of the objective, when it was illuminated by searchlights. False radio signals sent out in German won a few minutes' delay, enabling the *Campbeltown* to pass the last point at which she could be fired on by the main coast batteries, so that she was exposed only to the fire of light weapons for the last 5–6 minutes of the run in. When the enemy's guns opened fire, all the guns on the attacking vessels replied. The *Campbeltown* moved at top speed to break through the torpedo net guarding

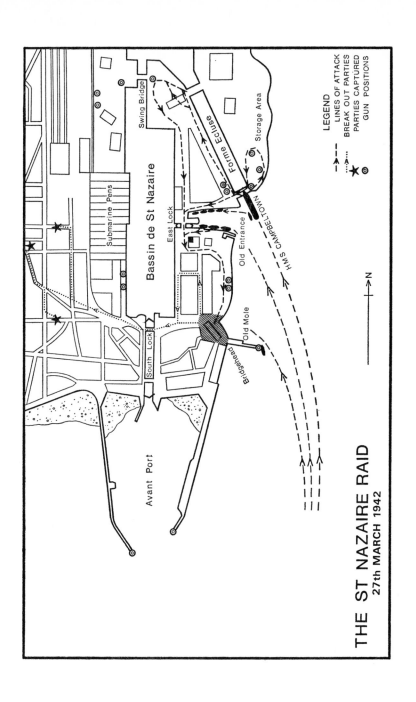

THE ST NAZAIRE RAID
27th MARCH 1942

the lock gate (of the Forme Ecluse) and crash into it with her bows stuck fast as she was scuttled and sank.

The assaulting groups were divided into Headquarters, Assault, Protection and Demolition parties, with a Special Task party and a reserve of 12 men. The job of the assault parties of 2 officers and 12 other ranks each, armed with Tommy-guns, Brens and rifles, was to form a bridgehead and perimeter, blocking all lines of approach from the main town; to clear the enemy from the outer harbour and destroy the guns there; eliminate the gun positions on each side of the main drydock entrance; to put the guns on the roof of the pumping station out of action; and to destroy the flak towers at the north end of the dock. The Special Task party was to destroy 2 guns between the Old Mole and the Old Entrance and to damage any ships they came across. The protection parties, armed like the assault parties, and consisting of 1 officer and 4 men each, were to guard the demolition parties, whose strength varied according to their task (their only personal arms were revolvers). Their role was to place charges below water level at the main gate to ensure its destruction after it had been rammed by the *Campbeltown*, to blow up the pumping station, to destroy both winding houses and to smash the inner drydock gate—if successful this would put the great drydock out of action for months. Then with only 1½ hours for their task, if time and circumstances permitted, they were to blow up the bridge connecting the dock area with the mainland and, when the troops had withdrawn to the old Mole (the place of reembarkation), to blow the bridge and dock gates connecting the Old Entrance with the Bassin St Nazaire to prevent any counterattack from making contact and also to make the entrance impossible for U-boats. Finally, the two bridges and sets of dock gates and the lock connecting the outer harbour with the Bassin St Nazaire were to be destroyed.

All but 1 of the 7 motor launches whose job was to land troops on the Old Mole were destroyed or disabled before they could do so. Of the 6 MLs that were to land their troops at the Old Entrance, 1 did so, and 2 others missed the entrance but regained their bearings, and turned to land their commandos. The remaining 3 MLs in this column were hit before landing their troops.

As soon as the *Campbeltown* struck, Colonel Newman, the Commanding Officer, and his party went ashore at the Old Entrance from motor gunboat 314, and made straight for the point selected as Headquarters, where they were to meet another party—which did not arrive. Soon the small group was under heavy fire, but was relieved by the arrival of a small group with a 2-inch mortar, which temporarily silenced the guns on the submarine pens.

That part of the force put ashore at the Old Mole encountered heavy and stubborn opposition from two defended pillboxes (which were never completely put out of action), from guns in a high building near the submarine pens, and from machine-guns mounted on the roofs of buildings. Heavy explosions indicated that the demolition parties were carrying out their tasks, although opposition was heavy and considerable fighting took place.

The commando group that went ashore from the *Campbeltown*, organised in a HQ and 4 parties, had a number of men wounded before getting ashore, but completed all its tasks successfully within the allotted time. Then it withdrew as planned to the bridge over the Old Entrance where, in due course, it was joined by the survivors of the parties which had got ashore at the Old Entrance and the Old Mole.

It was obviously impossible to disembark from the Old Mole, so Colonel Newman and about 50 men, many slightly wounded, after grouping at Headquarters, split up and tried to make their way through the town into the open country, intending to return to England through France and Spain. In the event the majority of them, including Newman (who was awarded the VC), were captured.

HMS Campbeltown blew up at 10.30 am, shattering the dock gate and so damaging the dock that it was never repaired by the Germans.

The British casualties were 144 killed or missing and 215 taken prisoner. The German casualties are believed to have been about 70 killed and an unknown number of wounded.

The battle can begin by involving only those troops who disembarked, accepting that those on vessels known to have been sunk took no part in the simulation. Another way is for the raiding force to be marked on the map at a specified point in the river, and then proceed, with vessels taking casualties from fire as they near the objective. Military Possibilities and Chance Cards can recreate the historical situation, in which the German coastal batteries opened up very late, so allowing the force to get inshore.

The targets of the original raid are listed, but the inclusion of all of them in the table-top reconstruction will be impossible. The targets chosen will be named on the commandos' map and the prime objective will be to destroy them within a specified time. The fate of the raiding party is a secondary consideration, nor is there any value in killing large numbers of Germans.

The 3–4,000 German soldiers were obviously not all in the immediate area of the docks at the onset of the raid; though sentries must have been on duty, pillboxes and other defence posts manned, and coastal batteries and flak towers obviously alerted by the earlier air raid, or from the moment firing first began in the river. Therefore, these posts can be manned by adequate numbers of Germans, while the remainder of the force is considered to be 'in barracks' somewhere in the town, so that, once alarmed, it will have to be assembled and transported to the area of fighting. On a Time Chart mark (a) the time at which the alarm reached them, (b) delay while they are rising, dressing, arming themselves and forming up, and (c) the time taken to embus or proceed on foot to the dock area, considered to be the German 'baseline'. Up to that point the Germans will be moving 'on paper', and they will only manoeuvre tactically, and as figures on the table, when coming forward from their baseline.

This aspect of the battle can be simulated by having a German 'controller' in another room, with an accurate scale map of the area, a Time Chart and a scaled table of movement. Fed with notes sent back from the points of conflict, he will act by sending troops wherever required.

Rating of Commanders and Observations

Colonel Newman and Commander Ryder, the Naval commander, will be 'above average' and the German commander, who apparently did nothing particularly wrong or inadequate, can be considered as 'average'.

The morale and fighting qualities of the raiders must always be higher than those of the defending Germans for the following reasons:

1. The attackers were selected, specially trained assault troops.
2. The defenders were run-of-the-mill garrison troops, who might even have been low category men invalided from the Russian Front.
3. The attackers had darkness and surprise as their very powerful allies.
4. The defenders were detrimentally affected by the darkness and surprise, which would make them apprehensive and lower their morale and fighting qualities.

These factors should have speeded up the raiders' reactions, so that, in any face-to-face confrontation, they are always assumed to have fired or taken offensive action before the Germans. The reconstruction will end if the raiding party can get a substantial proportion of its party, with wounded etc, taken off from the Old Mole; or at a point where all its possible targets have been eliminated; or when it is so decimated that few if any of the targets can be destroyed.

There is scope in this operation for some ingenious rules concerning the simulation of explosives and their effectiveness; there should be specified time-factors allowing men to be clear of a building before the charge explodes.

The chapter called 'Personalised Wargames' in *Advanced Wargames* sets out a method of carrying out a commando raid

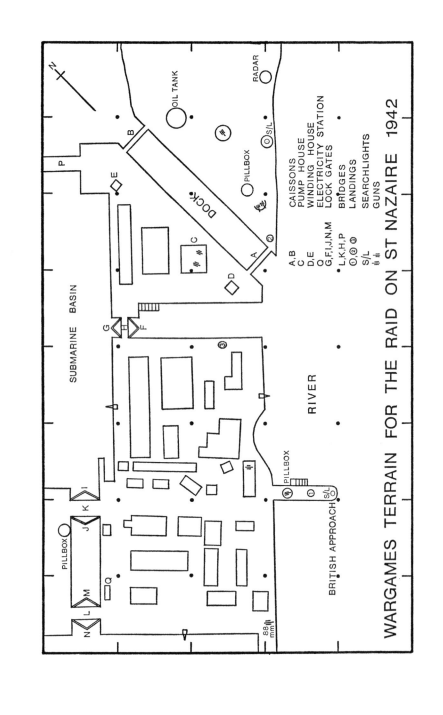

WARGAMES TERRAIN FOR THE RAID ON ST NAZAIRE 1942

A,B   CAISSONS
C   PUMP HOUSE
D,E   WINDING HOUSE
Q   ELECTRICITY STATION
G,F,I,J,N,M LOCK GATES
L,K,H,P   BRIDGES
①,②,③   LANDINGS
S/L   SEARCHLIGHTS
           GUNS

N

SUBMARINE BASIN

OIL TANK

RADAR

PILLBOX

DOCK

RIVER

PILLBOX

PILLBOX

BRITISH APPROACH

88 mm

similar to this, with the wargamer naming each member of his force.

Construction of Terrain

It is not one that can be knocked up in an odd hour before the game starts, because it requires buildings, docks, etc. The dock area can be chalked out on a wargames table or represented by a sheet of card, hardboard or plywood, with the outlines of the docks and jetties, etc, either cut out or marked in. The buildings can be made from card or blocks of wood; or there are excellent plastic kits of factories and buildings used in model railway construction. The destroyer *Campbeltown*, or destroyers like her, exist as kits; the other vessels can be scratch-built, made up from kits or card, or hardboard shapes can be used. The soldiers themselves are cheaply and readily available from sets put out by Airfix.

15
The Attack on Pork Chop Hill
16/18 April 1953

THIS SMALL-SCALE action was fought between United States infantry and a Chinese Communist force in Korea where, in late 1952 and early 1953, to gain psychological advantages at the Panmunjom Armistice negotiations, the Chinese Communist forces were attempting to overrun UN outposts. Pork Chop Hill, dominated by Chinese-held ridges and inadequately supported by neighbouring positions, was one such outpost. As Vietnam has shown, Korea set the pattern for future wars, with relatively 'soft' soldiers of the Western Democracies battling against hardy fanatical Communist peasants.

On the night of 16 April 1953 Pork Chop Hill was garrisoned by 76 men of the 1st and 3rd Rifle Platoons, Company E, 31st US Infantry Regiment, commanded by Lt T. V. Harrold. Paired-off, 20 GIs and Koreans formed outguards in 10 listening posts sited crescentically around the lower forward and flank slopes. Another group of 5 men formed half a patrol prowling the small valley forward of the hill. The hilltop itself was encircled by a solidly revetted rifle trench with some roof protection, so that it could be defended against attack from any direction. At 30 yd intervals, loopholed bunkers, sandbagged and heavily timbered, formed part of the trench line. At the rear, the top of the hill was pushed in like the dent in the top of a hat, forming a divided perimeter that forced the two defending rifle platoons into separate compartments, loosely joined in the centre.

Undetected by the outlying patrols, at about 10.30 pm, 2

companies of CCF infantry, each of about 70 men formed in 3 platoon columns, crossed the valley almost to the UN position without being seen. When sighted by an outpost, they raced forward and were in the trenches before a shot was fired. Surging like a flood over 1st Platoon's section, they machine-gunned or grenaded the bunkers, killing all but 7 of its 20 men. The Communists had hit the position at both ends of 1st Platoon's sector on the left end of the hill and, once in the trenches, they pinched in towards each other.

The Korean confrontation was a conflict fought largely with surplus weapons from World War II, and no startling tactical developments came out of it. By 1953 the best UN soldiers had either been killed or gone home on rotation, and the replacements were green soldiers, unhappy at fighting an unpopular war. The Chinese soldiers, harshly disciplined peasant conscripts, were furtive, fast and skilled, able to move stealthily in the deep valleys and steep hills of Korea. Their night attacks on the heavily defended UN outposts were often successful within minutes. However, they rarely held them under the quickly massed superior fire of American artillery, and the heavy attacks launched by day with artillery, air and armour support.

At 11 pm Lieutenant Harrold fired a red flare signifying that he was under attack and requesting artillery support; and at 11.05 both UN and Chinese guns laid a barrage around the base of the hill. For most of its length on the west side of the hill, where the attack came in, the trench had been covered over with pinewood beams, heaped with sandbags and, in some places, several feet of earth. Crushed in by artillery fire at several points, the fallen roof formed barricades that a rise in the trench-floor turned into successive terraced strongpoints. Here and in other parts of the hilltop position, CCF and American infantry fought with grenades and sub-machine guns in the maze of trenches and bunkers, but the hill was overrun and, although at dawn the Americans were still holding bunkers and isolated strongholds, for practical purposes the outpost had been lost, since the CCF had gained possession of more than half the hill. Either through exhaustion, fear, the incoming barrage from both UN and CCF

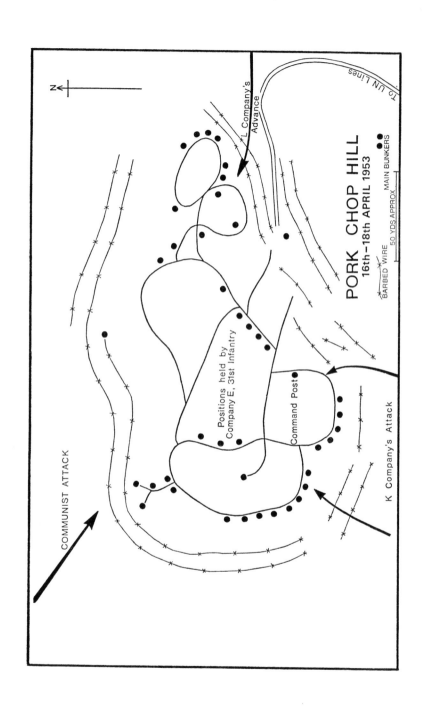

PORK CHOP HILL
16th–18th APRIL 1953

MAIN BUNKERS

BARBED WIRE

50 YDS APPROX

To UN Lines

L Company's Advance

N

COMMUNIST ATTACK

Positions held by Company E, 31st Infantry

Command Post

K Company's Attack

artillery, or insufficient ammunition (carrying little, they relied on captured ammunition), the Chinese infantry squatted in and on the bunkers instead of continuing down the trench line and mopping up.

The defence put up by 3 officers and 4 NCOs in the Command Post on the far side of the position from the CCF attack, made the bunker a blockhouse that barred the rear door to the hill, preventing the Chinese from gaining the rear slope and cutting off both line of retreat and route for reinforcements. In the early hours of the morning, 2 platoons were sent up the hill to reinforce the garrison from F and L Company, but one got lost in the dark and did not arrive, while the other, coming unexpectedly under Chinese fire, ran back into the valley. Then, at 3.30 am Lt Clemons, commanding K Company (135 men), was ordered to assault Pork Chop from its rear while two platoons from L Company went up the hill from the right. K Company riflemen each carried a full belt and bandolier of ammunition for their Garand rifles, and 3 or more grenades. There were 6 Browning Automatic rifles firing 500 ·30 rounds per minute with 12 magazines per weapon. Each platoon carried a flamethrower and a 3·5 mm rocket launcher (Bazooka).

K Company advanced up the hill, deploying its 2nd platoon on the right, 1st on the left and 3rd in reserve. From the assault line at the foot of the hill it was only 170 yd to the nearest bunkers, but the route was very steep, strung with wire and cratered, with tree stumps and rocky outcrops, so that it took 30 minutes unopposed climbing to reach the top. The Chinese had burrowed into the hill and their artillery doused the area in regular 10 minute timed patterns. K Company fought for 2 hours, becoming exhausted as it gained 200 yd in a series of individual actions by small groups—all group-initiative having vanished.

The platoons from L Company advancing on a narrow front on the right were mown down by the entrenched defenders so that of the 62 that had attacked under Lt Crittenden, only 10 exhausted men arrived to join Clemons. At 8 am, with both Americans and Chinese out of the trenches and grouped on higher ground inside the parados, so that they fought at a few feet range, K Company were out of water and short of ammunition. They did not know that the enemy was in a similar position and a determined attack might have recaptured the hill. At about this time, under heavy artillery fire, G Company of the 17th Infantry Regiment, under Lt Russell, struggled up the back slope and joined Clemons, whose force consisted of 35 survivors of K Company, 10 men from L Company and 12 men from Harrold's E Company, rescued from the rubble.

At the same time as G Company was creeping up the cratered slopes, a fresh Chinese company pushed on to the other end of the ridge. The battle suddenly blazed up as men fired and threw grenades amid the jumble of tumbled trenches, shattered bunkers and shell holes, under an almost unceasing artillery fire that fell on the hill. Soon the newly arrived G Company was down to 50 men.

During the early part of 17 April, Clemons was using three Sherman tanks, positioned 500 yd behind the hill, as artillery to deter the CCF from approaching over low ground west of the hill but, at midday, they obtained permission to depart, fearing that a CCF success on the hill would make them a target for Communist artillery. Then the survivors of G Company were ordered by a misinformed HQ to leave the hill, so at 3 pm Russell had to pull out, leaving Clemons with only 25 men.

While this handful of exhausted men held on to the hill, because the fight was local but the issue national, Division, Corps, Army and Far East Command Headquarters were all debating the simple question: 'Do we really want to hold Pork Chop Hill?' If the hill were yielded, the Communists would strike at another position, but if it were fought for, it might become an attritional trap for battalion after battalion. Finally, UN accepted that it was going to have to stand up to the CCF or be at a serious disadvantage in the armistice negotiations at Panmunjom, so Lt Denton took L Company on to Pork Chop to join Clemons' remaining 16 men and, at 9 pm F Company mounted the back slopes and relieved Clemons.

But under the desperate Chinese artillery pounding and infantry attacks, this reinforcement was found to be insufficient, and E Company of the 17th Infantry had to be committed; while at dawn A Company was also sent in as company after company of reinforcing Chinese mounted the hill. Throughout the night and all day of 18 April the UN force, taking 60 and 75 per cent casualties, fought on, until finally the CCF realised that they were not to have the hill, and just after sunset on 18 April the assault ceased as quickly as it began. In his book *Pork Chop Hill* S. L. A. Marshall wrote: 'All the heroism and all of the sacrifice went unreported. So the very fine victory of Pork Chop Hill deserves a description of the Won-Lost Battle. It was won by the troops and lost by the people who had sent them forth'.

Using Airfix model soldiers this action can be fought on a man-for-man basis without scaling-down numerical strengths.

Rating of Commanders and Observations

Lt Harrold may be classed as 'average', Lt Clemons as 'above average', and the Chinese commanders as 'average'.

Certainly at the start of the action the CCF would seem to be entitled to a higher morale state, perhaps declining as the action progresses. UN morale is more variable, reflected by local rules that allow the Americans momentarily to break but quickly to recover and fight on. If the Americans are to hold, as they did historically, in spite of the numerical disparity, both sides must be given equal fighting qualities, though at this stage of an unpopular war, with most of the best men gone, it is reasonable to assume that the young American conscripts, fighting in alien territory, were not as effective as their Chinese peasant opponents. Morale differences are balanced by the superior weapons and plentiful ammunition of the Americans. The Chinese, in fact, might well be subject to a Military Possibility that causes them to run out of ammunition and to be unable to find a supply nearby.

At the start of the action E Company is in its defensive positions on and around Pork Chop Hill; both commanders mark their dispositions and movements on a map of the hill and surrounding terrain. Ideally the wargamer representing

K

Lt Harrold, E Company commander, should be unaware of the direction of the CCF attack, which, in fact, reached the position without being seen. The umpire will plot on his master map the progress of the attacking Chinese and whether or not they stumble upon an American/Korean patrol. A Military Possibility can give the American patrols a slight chance of raising the alarm, but the short period of time between the warning and the onslaught will not allow much to be done. The surprise element of the CCF attack can be represented by giving them first fire and imposing a morale penalty on the Americans.

Historically, Lt Harrold called for artillery support, and a Military Possibility (by the use of Chance Cards, etc) can decide whether or not he has time to do this, though the odds should be in his favour. The call for support must be plotted on a Time Chart, which will show that at 11.05 pm the American artillery attempted to seal off the CCF end of the hill, and the Chinese artillery soon tried to do the same at the American end; both sides fire 'off-table map-shoots'. The Time Chart must also plot the gunfire of both sides, which frequently halted the fighting for intervals that must be specified on the Time Chart. During these pauses both sides sheltered in bunkers and in the covered part of the trench. In fact, at about 2 am there were enough Chinese on Pork Chop to mop up and capture it, but the UN and CCF artillery fire prevented them from venturing into the open. These factors must be simulated during the reconstruction. Chinese artillery fired at approximately 10 minute intervals almost throughout the entire action; this can be marked on the Time Chart, or the CCF can do a map-shoot at the conclusion of each game-move.

The Chinese soldiers were more at home in the darkness than their American opponents, which advantage may be simulated by throwing a dice whenever there is a confrontation between individuals or parties:

4, 5 or 6—the Chinese fire first,
2, 3—the Americans fire first,
1—they fire simultaneously.

Using the Time Chart, bring Clemons and K Company on to the table at the rear of the hill, with the two platoons from L Company coming on the right. If the reconstruction has progressed so as to allow Chinese soldiers to oppose them (in fact, K Company climbed unopposed but L Company took heavy losses) Military Possibilities can determine whether the CCF opposes both, opposes neither, opposes K or L individually, or causes greater or fewer casualties than were caused in the action itself.

Before this, two platoons of F and L Companies had been sent as reinforcements, but neither arrived—a Military Possibility could make one or both of them get on to the hill. Similar measures could affect Russell's G Company coming on to the hill and withdrawing at 3 pm, and movements of the various Chinese reinforcing groups. The historical times of these arrivals and departures should be marked on the Time Chart, and then Military Possibilities should be used to ascertain whether they are carried out as they occurred or whether they deviate.

It is recorded history that it was decided to reinforce Clemons and, if our simulation is to be accurate, this must occur, but Military Possibilities can hasten or slow down the eventual arrival of Lt Denton's L Company, and then F Company and others. The 3 UN tanks at the foot of the hill could, by a Military Possibility, be called on to the hill by Clemons, where the jumbled terrain would make them extremely vulnerable, though this destroys much of the credibility of the simulation.

This is a highly suitable action for the 'personalised' style of wargaming. Marshall's excellent book includes many of the names of the combatants.

Construction of Terrain

As the battle takes place wholly on the hilltop, it can form the entire wargames table as an undulating, cratered and very jumbled terrain. The approach marches by both Americans and Chinese will be carried out on the map, the forces being placed on the table at their points and times of arrival. A sand table is the ideal medium for this terrain, though a satisfactory substi-

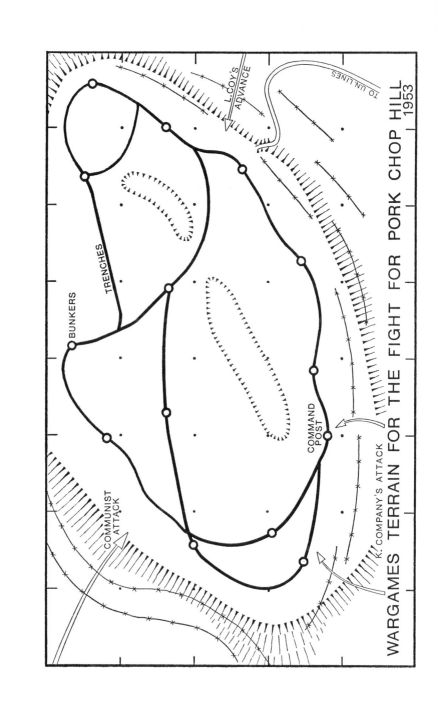

WARGAMES TERRAIN FOR THE FIGHT FOR PORK CHOP HILL 1953

TO UN LINES

L COY'S ADVANCE

TRENCHES

BUNKERS

COMMAND POST

K COMPANY'S ATTACK

COMMUNIST ATTACK

tute is a cloth draped over 'hill' shapes, with trenches and bunkers made of plasticine, balsa-wood, etc. A good selection of items like bunkers, trenches, etc, are made by Bellona, and they can easily be built into the terrain. A semi-permanent terrain can be made out of hessian or sacking soaked in polyfilla and allowed to dry in premoulded shapes; it can be used for any number of similar actions in World Wars I and II, Korea or Vietnam.

Appendix 1
Rules

RULES ARE principles to which actions or procedures conform or are bound or intended to conform. In wargaming they are based on theory plus the experience and practice of the players themselves who devise them to cover the procedure of their table-top battles, with the intention of bringing to the battle-simulation the greatest possible realism. The majority of readers will have rules of their own that will suffice to control the battles in this book, when supplemented by the suggestions contained in these pages.

There are sets of commercially produced rules, such as those of the Wargames Research Group (75 Ardingly Drive, Goring-by-Sea, Sussex), which publishes a set of Ancient rules 1000 BC to 1000 AD that will cover Pharsalus and Poitiers and probably Barnet; its Horse and Musket Rules 1750–1850 will cover Prestonpans, Maida, Aliwal and perhaps Wilson's Creek; and its Infantry Platoon Action Rules will control the St Nazaire raid and Pork Chop Hill and, at a pinch could be adapted to fit the Gallipoli battle. Cheriton is catered for by rules for the period 1500–1600 produced by D. Millward (36 Bells Lane, Kings Norton, Birmingham B14 5QN) and English Civil War Rules from D. Featherstone (69 Hill Lane, Southampton SO1 5AD). From the latter source can be obtained sets of Ancient rules that will cover Pharsalus; Mediaeval Rules for Poitiers and Barnet; Eighteenth-Century Rules for Wynendael, Prestonpans and Guilford Courthouse; Napoleonic Rules for Maida; Horse and Musket Rules for Aliwal, Little Big Horn and (with adaptations) Modder River; American Civil War Rules for Wilson's Creek; World War I Rules for Gallipoli; and World War II Rules for St Nazaire and Pork Chop Hill. Maida is also very satisfactorily covered by the Napoleonic Wargames Rules of S. and J. Reed (33 Salvington Hill, High Salvington, Worthing BN13 3BD).

Most historical battles include numerical disparities, surprise factors, varying qualities of troops, inequalities of weapons and morale effects that are extremely difficult to handle under normal wargames rules. For example, the Battle of Aliwal is a particularly good example of a situation where the larger force, both in men and guns, was also protected by field entrenchments, and yet was completely routed by Sir Harry Smith's much smaller army. There are occasions in battle when a commander takes a deliberate calculated risk that no normal set of wargames rules would allow. Yet the history of warfare demonstrates that many such deviations from the rules are successful, and allowances should be made for breaking the rules on the wargames table.

This leads to the point that local rules are required on the wargames table. It cannot of course be denied that no known set of rules, lacking 'local' adaptations, would give victory to Clive at Plassey, to the British forces in almost any battle of the Indian Mutiny, or in at least half of the battles described in this book. 'Local' rules should therefore make allowances for factors peculiar to a specific battle. No Wargames Rules would allow the 2nd Guards at Guilford Courthouse to recover their morale, for they were completely routed by the 2nd Maryland Regiment and a cavalry attack in their rear, and then heavily hit when Cornwallis ordered his artillery to fire grapeshot into the mêlée. But they rallied in the shelter of the wood, joined the 1st Battalion of Guards and went forward again to the attack. Local rules only can cater for these exceptional incidents.

Eleven out of the 15 battles herein involve troops of varying qualities, such as Greene's blend of untrained Militia and Continental Regulars at Guilford Courthouse. At least 6 battles demonstrate marked inequality of weapons and weapon handling, such as the English longbowmen at Poitiers and the American use of flame-throwers on Pork Chop Hill. Distrust of their comrades was shown by the Lancastrians at Barnet and by the Transvaalers at Modder River. The weather played a decisive part at Barnet and, to a certain degree, affected the Cheriton battle; while darkness played a marked role at Prestonpans, Wilson's Creek, St Nazaire and Pork Chop Hill. In no less than 12 battles some or all of the troops were materially affected by morale conditions perhaps peculiar to that conflict.

All these factors have to be taken into account if the reconstructions of the battles in this book are to be more than giving names to wargames.

Appendix 2
Terrain

AT LEAST three-quarters of the battles described in this book are vitally affected by the terrain over which they were fought—the positions at Poitiers and Wynendael played a big part in allowing a small force to defeat a numerically superior army; the 'Bloody' lane at Cheriton caused the downfall of the Royalist cavalry; while the maze of gulleys, trenches and mountainous ground caused untold difficulties to the troops at Gallipoli and on Pork Chop Hill.

When attempting to reconstruct any real-life battle, the terrain is perhaps the most important factor in the project because, both in topographical features and dimensions, it must closely resemble the actual battlefield, otherwise what takes place upon it will bear only the most coincidental resemblance to the historical events under simulation.

With 3 exceptions, the confrontations covered by this book can be constructed within the bounds of a playing surface of 8ft × 5ft. The exceptions are the Little Big Horn, Wilson's Creek and Modder River, where the relatively extensive battle areas lend themselves to division into separate actions, with a coordinated end-result.

When reconstructing a historical battlefield there are two important considerations:

1. It is only necessary to reproduce those areas of the battlefield over which combat took place, so that all possible space on the wargames table can be utilised.

2. Even though it might mean 'ironing out' the known

contours of the actual battlefield, all hills and slopes must be so angled as to allow model soldiers to stand up on them.

Perhaps the most realistic wargames terrain is a sand table moulded into hills, valleys, sunken roads, river beds, trenches, shell holes, etc. Sand-table terrains are considered in some detail in my book *Wargames* and my booklet *Wargames Terrain.*

A method of simulating hills, valleys and undulating ground is to stretch a green cloth or plastic sheet over carefully assembled mounds of books, slabs of polystyrene or pieces of wood. Rivers and roads can be painted with poster paint on to the sheeting and look extremely realistic. Slabs of wood or polystyrene placed upon each other to make 'stepped' hills give the terrain a 'symbolic' appearance, providing readily definable contours plus an ideal surface for soldiers to stand upon.

Trees can be made from lichen moss stuck on to pieces of twig; hedges from the same material. Stone walls and rail fencing can be bought or made from balsa wood or plastic ceiling tiles. In fact, these ceiling tiles can be used for houses, bridges, castles, and, when suitably roughened up and painted, make excellent crags and rocky outcrops.

There are innumerable plastic kits of houses, factories, etc, on the market, but the relatively specialised type of buildings required for the St Nazaire terrain may make it necessary for the wargamer to scratch-build.

Although vegetation need not be portrayed on the terrain maps, few stretches of land are completely bald, and the appearance of any battlefield is greatly improved by scattered clumps of trees, bushes and scrub. The terrains of Cheriton, Wynendael and Wilson's Creek had numerous wooded areas. Wargames table woods must be open for troops to manoeuvre without knocking over the trees, so use irregularly shaped pieces of hardboard, painted dark green, with 3–4 trees around the perimeter.

Bellona Battle Game scenery and landscape models (produced by Micro-Mold Plastic, 1 Unifax, Woods Way, Goring-by-Sea,

Sussex, England) provide an assortment of earth brown or sand-coloured PVC trenches, sandbagged emplacements, bunkers, ruined buildings, pillboxes, encampments and revetted earthworks. Quickly coloured, the resulting scenic setpieces are most realistic and effective, and give a professional effect to the terrain.

Appendix 3
Availability of Model Soldiers

It is doubtful if any of the commercial model soldiers manufacturers listed below could supply figures for all 15 battles in
this book. Nor is it certain that, apart from flat figures, there are
available ranges for all the participants, namely the Boers at
Modder River, the Anzacs and Turks at Gallipoli, and the CCF
soldiers who attacked Pork Chop Hill. However, the industrious
wargamer can fairly easily convert existing Airfix figures into
the missing troops. There are financial limitations to the number
of troops a wargamer can amass for any particular battle or
campaign, metal figures costing 7–10 p for infantrymen and about
double that price for cavalry. The armies required for a
reconstruction of the Battle of Poitiers, for example, would cost
between £60 and £90, although they could be collected on a far
cheaper basis if Airfix plastic figures from the Sheriff of
Nottingham and Robin Hood sets are used; but there is a vast
amount of painting and converting to be done before the
colourful forces face each other across the battlefield.

The constantly increasing range of 20 mm plastic figures
manufactured by Airfix Hobby and Toy Sales Ltd are easily
convertible to *any* desired soldier of *any* period. This fact puts
wargaming in any period within the reach of even the schoolboy
with limited financial resources. For instructions on converting
Airfix plastic figures, read Chapter 5 of my book *Military
Modelling* (1970). *Airfix Magazine* and *Wargamers' Newsletter*
run regular illustrated articles on plastic figure conversion.

Wargames figures are now obtainable in the following sizes:
5 mm; 9 mm; 15 mm; 20 mm; 25 mm; 30 mm; 35 mm; and
40 mm. Each manufacturer has his own conception of scale, so

it is best to check on figure sizes before buying different makes. The principal manufacturers throughout the world are listed below. For more detailed information on all known manufacturers of model soldiers, see my volume *Handbook for Model Soldier Collectors* (1969).

Airfix Hobby & Toy Sales Ltd (plastic 20 mm). From Woolworth's and hobby shops.

Douglas Miniatures (metal 20 mm), 17 Jubilee Drive, Glenfield, Leicester, LE3 8LJ.

The Garrison (metal 20 mm), 198 Northolt Road, South Harrow, Middlesex, MA2 0EN.

Frank Hinchliffe (metal 25 mm), 83 Wessendenhead Road, Meltham, Yorks.

Hinton Hunt Figures (metal 20 mm), Rowsley, River Road, Taplow, Bucks.

P. Laing (15 mm, Marlburian only), 11 Bounds Oak Way, Southborough, Tunbridge Wells, Kent.

W. H. Lamming (metal 20 mm), 130 Wexford Avenue, Greatfield, Kingston-upon-Hull, Yorks.

Miniature Figurines (metal 5 mm, 15 mm, 25 mm and 30 mm), 28/32 Northam Road, Southampton, or 100A St Mary Street, Southampton SO1 1PB. MF have biggest range and variety of figures.

The Northern Garrison (20 mm and 30 mm metal), Castlegate, Knaresborough, Yorks.

Oscar Figures (metal 30 mm), Longcroft, The Green, Little Horwood, Bletchley, Bucks.

Phoenix Model Developments Ltd (metal 20 mm, 25 mm and 30 mm), The Square, Earls Barton, Northampton.

Rose Miniatures (metal 20 mm), 45 Sundorne Road, Charlton, London SE7.

R. W. Spencer-Smith (30 mm plastic), 66 Longmeadow, Frimley, Camberley, Surrey.

Springwood Models (20 mm Napoleonics, plastic), Oxford Model Centre, 94 St Clements, Oxford.

Tradition (metal 25 mm and 30 mm), 188 Piccadilly, London W1.

Warrior (metal 25 mm and 30 mm), 23 Grove Road, Leighton Buzzard, Beds, LU7 8SF.

Willie Figures (metal 30 mm), 60 Lower Sloane Street, London SW3. Sikh Wars, among others.

USA

Bugle & Guidon (metal 30 mm), PO Box 662, Jackson, Mich 49204, USA. Custer's 7th Cavalry and Indians.

Command Post (metal 30 mm), 760 West Emerson, Upland, California 91786, USA.

Hobby House (metal 20 mm), 9622 Ft Meade Road, Laurel, Maryland 20810, USA.

C. H. Johnson (metal 20 mm, 25 mm and 30 mm, mostly British imported), PO Box 281, Asbury Park, New Jersey 07713, USA.

Jack Scruby (metal 9 mm, 20 mm, 25 mm, 30 mm and 40 mm), PO Box 3088, Visalia, California 93277, USA.

K. & L. Company (metal 20 mm, American Civil War only), 1929 North Beard, Shawnee, Oka 74801, USA.

Der Kriegspeiler (20 mm metal, Napoleonics only), PO Box 419, Bedford, Mass, USA.

SPAIN
Alymers Miniploms (metal 20 mm), Maestro Lope 7, Burjasot, Spain.

SWEDEN
Holgar Eriksson (metal 30 mm), Sommarrovagen 8, Karlstad, Sweden.

FRANCE
Segom (plastic 30 mm Napoleonics), 50 Boulevard Malesherbes, Paris 8, France and Model Figures and Hobbies, 8 College Square North, Belfast BT1 6AS, Ireland.

Starlux (plastic 30 mm), UK distributors – Beatties Ltd, 112 High Holborn, London WC1.

WEST GERMANY
Elastolin (plastic 40 mm) Neustradt/Coburg, Jahreswende, West Germany.

Flat Figures are two-dimensional 30 mm figures stamped out of thin metal. It is possible to obtain *any* type of soldier, of *any* period of history and in *any* position from one or other of these suppliers:

Aloys Ochel, Feldstrasse, 24b, Kiel, West Germany.

Rudolf Donath, Schliessfath 18, Simbach/Inn OBB, West Germany.

F. Neckel, Goethestrasse 16, Wendlingen Am Neckar, West Germany.

Bibliography

In ADDITION to those listed below, the following may also be consulted:

Dupuy, R E. and T. N. *The Encyclopedia of Military History from 3500 BC to the Present* (1970): Pharsalus, Poitiers, Barnet, Prestonpans, Guilford Courthouse, Maida, Wilson's Creek, Little Big Horn, Modder River, Gallipoli, St Nazaire

Eggenberger, David. *A Dictionary of Battles from 1479 BC to the Present* (1967): Pharsalus, Poitiers, Barnet, Prestonpans, Guilford Courthouse, Maida, Wilson's Creek, Little Big Horn, Modder River, Gallipoli

Kinross, John. *Discovering Battlefields in Southern England* (paperback, nd): Barnet, Cheriton, Prestonpans.

PHARSALUS

Hadas, Moses and the Editors of Time-Life Books. *Imperial Rome* (1966)

Montgomery, Field-Marshal Viscount. *A History of Warfare* (1968)

Webster, Graham. *The Roman Imperial Army* (1969)

POITIERS

Bryant, Arthur. *The Age of Chivalry* (1963)

Burne, Alfred H. *The Crécy War* (1955)

Featherstone, D. F. *The Bowmen of England* (1967)

Montgomery, Field-Marshal Viscount. *A History of Warfare* (1968)

BARNET

Barnett, C. R. B. *Battles and Battlefields in England* (1896)

Fortescue, Sir John W. *A History of the British Army*, Vol I (1899)

Grant, James. *British Battles on Land and Sea* Vol I (nd)

CHERITON

Adair, John. *Cheriton* (1972)

Godwin, G. N. *The Civil War in Hampshire* (1904)

Rogers, Colonel H. C. B., OBE. *Battles and Generals of the Civil Wars 1642–1651* (1968)

WYNENDAEL

Atkinson, C. T. *Marlborough and the Rise of the British Army* (1921)

Belfield, Eversley. *Oudenarde 1708* (1972)

Fortescue, Sir John W. *History of the British Army*, Vol I (1899)

Grant, James. *British Battles on Land and Sea* (nd)

PRESTONPANS

Fortescue, Sir John W. *History of the British Army*, Vol II (1910)

Tomasson, Katherine and Buist, Francis. *Battles of the '45*, (paperback, 1962)

GUILFORD COURTHOUSE

Fortescue, Sir John W. *History of the British Army*, Vol III (1911)

Reid, Courtland T. *Guilford Courthouse*, National Park Service Historical Handbook Series No 30 (paperback, 1959)

MAIDA

Battles of the Nineteenth Century Described by Archibald Forbes, G. A. Henty, Major Arthur Griffiths and Other Well-Known Writers, Vol II (1902)

Fortescue, Sir John W. *History of the British Army*, Vol V

ALIWAL

Battles of the Nineteenth Century, Vol II (1902)
Featherstone, D. F. *All For a Shilling a Day* (1966)
Featherstone, D. F. *At Them With the Bayonet!* (1969)
Fortescue, Sir John W. *History of the British Army*, Vol XII (1934)
Lunt, Sir J. *Charge to Glory* (1962)
Moore-Smith, G. C. *The Autobiography of Sir Harry Smith* (1901)

WILSON'S CREEK

Catton, B. *This Hallowed Ground* (1957)
Commager, H. S. *The Blue and the Grey*, Vol I (1950)
Johnson, R. V. and Buel, C. C. (eds). *Battles and Leaders of the American Civil War*, Vol. I. (1956)
Pratt, F. *Ordeal by Fire* (1950)

LITTLE BIG HORN

Battles of the Nineteenth Century, Vol I (1902)
Luce, Edward S. and Evelyn S. *Custer Battlefield*. Historical Handbook Series, No 1 (paperback, 1949)

MODDER RIVER

Battles of the Nineteenth Century, Vol III (1902)
Pemberton, W. Baring. *Battles of the Boer War* (paperback, 1964)
Wilson, H. W. *With the Flag to Pretoria: A History of the Boer War of 1899–1900*, Vol I (1900)

ANZAC LANDING AT GALLIPOLI

Armstrong, H. C. *Grey Wolf* (paperback, 1935)
James, Robert Rhodes. *Gallipoli* (1965)
Soldiers Battle Tales from 'Blackwood's' (1968)

ST NAZAIRE

St Nazaire Raid, The (Campaign Book), Purnell's History of the 2nd World War (paperback)

Saunders, Hilary St George. *The Green Beret* (paperback, 1949)

PORK CHOP HILL

Fehrenbach, T. R. *This Kind of War* (1963)
Marshall, S. L. A. *Pork Chop Hill* (paperback, 1956)

Among the best-known books on wargaming are:
Little Wars, H. G. Wells (1913)
Wargames, Donald Featherstone (1962)
Wargames in Miniature, J. Morschauser (1963)
Naval Wargames, Donald Featherstone (1966)
Air Wargames, Donald Featherstone (1967)
Charge!, Brigadier P. Young and Lt-Colonel J. P. Lawford (1967)
Advanced Wargames, Donald Featherstone (1969)
Discovering Wargaming, John Tunstill (1969)
Introduction to Battle Gaming, Terence Wise (1969)
Battle! Practical Wargaming (*World War II*), Charles Grant (1970)
Battles With Model Soldiers, Donald Featherstone (1970)
Wargames Campaigns, Donald Featherstone (1970)
The Wargame, Charles Grant (1971)
Solo Wargames, Donald Featherstone (1972)
The Wargame, ed Peter Young (1972)
Wargames Through the Ages, Vol I: Ancient and Medieval Periods, Donald Featherstone (1972)

There are a number of regularly published magazines and journals dealing with wargaming, perhaps the best known and oldest being *Wargamers' Newsletter* (monthly, 69 Hill Lane, Southampton, SO1 5AD, England). Other publications in Great Britain and USA are the following:
The Armchair General—PO Box 274, Beltsville, Maryland 20705, USA
The Bulletin—The British Model Soldier Society, 16 Charlton Road, Kenton, Middlesex

The Courier—Bulletin of the New England Wargamers Association, 45 Willow Street, Brockton, Mass, USA

International Wargamer—3919 West 68th Street, Chicago, Illinois 60629, USA

Panzerfaust—PO Box 1123, Evansville, Indiana 47713, USA

Slingshot—Official Journal of the Society of Ancients, 757 Pershore Road, Selly Park, Birmingham 29

Strategy and Tactics—Simulations Publications Inc, 34 East 23rd Street, New York, NY 10010, USA

Tradition—The Journal of the International Society of Military Collectors, 188 Piccadilly, London W1

Index